ELECTRICAL ENGINEERING DEVELOPMENTS

CASE-BASED REASONING

STRATEGIES, DEVELOPMENTS AND APPLICATIONS

ELECTRICAL ENGINEERING DEVELOPMENTS

Additional books in this series can be found on Nova's website
under the Series tab.

Additional e-books in this series can be found on Nova's website
under the e-books tab.

ELECTRICAL ENGINEERING DEVELOPMENTS

CASE-BASED REASONING

STRATEGIES, DEVELOPMENTS AND APPLICATIONS

TAO LIN
EDITOR

nova publishers
New York

Library of Congress Cataloging-in-Publication Data

Case-based reasoning (Nova Science Publishers : 2015)
Case-based reasoning : strategies, developments and applications / editor, Tao Lin (director, Power System Automation Research Institute, assistant dean, School of Electrical Engineering, Wuhan University, Wuhan, Hubei province, China).
pages cm
Includes index.
ISBN 978-1-63482-952-6 (hardcover)
1. Systems engineering--Data processing. 2. Electric power systems--Management. 3. Machinery--Maintenance and repair. 4. Case-based reasoning--Industrial applications. I. Lin, Tao (Power system engineer), editor. II. Title.
TA168.C356 2015
620'.0042--dc23
2015031078

Published by Nova Science Publishers, Inc. † New York

CONTENTS

PREFACE

Case based reasoning (CBR), is a new analogical reasoning method in the artificial intelligence field, accompanied with the advance of cognitive psychology. The premise of CBR is that similar situations can lead to similar results and solved by similar methods. In detail, based on the structurally stored past cases (experiences and knowledge), using similarity as the measuring standard, the case which is most similar to the case/problem/situation to be analyzed currently can be found out as the reference solution.

Firstly proposed by Roger Schank in early 1980s and derived by numerous reseachers for more than 30 years, CBR has achieved great developments and wide applications. Compared with classical rule based reasoning (RBR), CBR has the following advantages:

1) Knowledge acquisition is easier. For CBR, its knowlege unit is case, which can be obtained without rigorous processes compared with the rule distillation of RBR.

2) Searching is more efficient. Different from the exact match of RBR for each rule, the match of CBR is imperfect matching based on the analysis of similarity, adopting the nearest neighbour approach, inductive approach, knowledge-based indexing method or template retrival approach.

3) Applicability is better. The case base of CBR can be continually modified and adjusted in the application, while RBR is lacks autonomous learning ability. Especially for large-scale complex problems, the application advantage of CBR will be more evident.

This book focuses on the application of CBR in multiple scientific fields, including power system transient stability assessment, power system small signal stability early warning, multiple power disturbance recognition, power transformer fault diagnosis, fault diagnosis and reliability prediction of high-speed railway, fault diagnosis system in machinery, and decision support for planning in the case of failures of technical systems and food science. As CBR has good applicability, the book is expected to contribute a better understading of CBR and how to suitably and fully apply the method in various realms not limited to the above subjects.

I would like to thank all of the authors for their wholehearted cooperation, active involvement and rigorous academic approach in compiling this book, and their contributions to making this book more comprehensive and meaningful. This book may have guiding significance for those who deal with resoning analysis in various kinds of scientific fields.

I am also grateful to the people related to the publication of this book for their help in collecting, sorting and editing this publication.

Tao Lin
School of Electrical Engineering Wuhan University,
Wuhan, China
Email: tlin@whu.edu.cn

In: Case-Based Reasoning
Editor: Tao Lin

ISBN: 978-1-63482-952-6

Chapter 1

A Reconciliation-driven Approach of Case-based Prediction: State of the Art, Method Overview and Application in Food Science

***Fatiha Saïs*[1]*and *Rallou Thomopoulos*[2]†**
[1]LRI Paris Sud University, Orsay, France
[2]INRIA Project Team GraphIK
and INRA Joint Research Unit IATE
Montpellier, France

Abstract

This chapter proposes an approach to generate predictions for decision support issues. It relies on case-based and reconciliation methods, using an ontology. The objective of the chapter is to provide an overview of the state of the art, but also to describe the proposed method and to illustrate it on a concrete application. In this approach, a reconciliation stage identifies groups of rules expressing a common experimental tendency. A prediction stage generates new rules, using both experimental tendencies obtained in the previous stage and new experimental descriptions. The method has been tested within a case study concerning food quality management. It has been compared to a classic predictive approach, leading to promising results in terms of accuracy, completeness and error rate.

Keywords: reasoning from knowledge, information integration, prediction, case-based reasoning, data reconciliation.

1. Introduction

Scientific publications in experimental sciences express syntheses of published experimental studies, obtained and validated through repeated experimentations. However, there is a

*E-mail address: Fatiha.Sais@lri.fr.
†E-mail address:Rallou.Thomopoulos@supagro.inra.fr.

huge number of possible experimental conditions. Only a limited part of the possible experimental conditions have been explored in the literature and established as domain knowledge. Therefore, to make predictions concerning unexplored experimental conditions, a solution consists in using existing knowledge that concern close – although not identical – experimental conditions. In the more classic case where one starts from raw data, this approach is the principle of case-based reasoning.

Experimental conditions sometimes only differ by a small variation of one experimental parameter, which may be fundamental in the case of a highly discriminant parameter, but negligible for others. Hence, knowledge that both (i) concern close experimental conditions and (ii) show similar results may be identified, which is a reconciliation problem. Such reconciled knowledge have a semantics, since they express a common experimental tendency. In addition to the experimental knowledge, general domain knowledge is available, and it has been modeled in an ontology. The ontology includes a vocabulary organized by subsumption, disjunction and synonymy relations. Moreover, it provides less common information concerning the status of concepts, such as functional dependencies and discriminance of concepts for prediction.

The objective of this chapter is to propose an approach to generate predictions relying on case-based and reconciliation methods, using an ontology. The approach has been tested within a food science application concerning food quality management in the cereal agri-food chain and it has been compared to a classic predictive technique.

The chapter is organized as follows: Section 2. gives an overview of related work in case-based reasoning, data reconciliation, as well as regarding the application domain. Section 3. provides the basic materials used for ontology and domain rule representation. Section 4. is dedicated to the proposed rule reconciliation method. Section 5. presents the proposed prediction method. Section 6. describes the context in the application domain and proposes a comparative evaluation of the developed approach. Finally, Section 7. concludes the chapter by giving some future work perspectives.

2. Related Work

This section presents previous works in case-based reasoning, data reconciliation (in the framework of information integration) and concerning the application domain.

2.1. Case-Based Reasoning

Case-based reasoning is a reasoning paradigm that relies on the adaptation of already solved problems, stored in a database, to solve new problems. This reasoning mechanism is not based on deduction, as rule-based reasoning, but on analogy [24].

A "case" is a description of a problem, associated with its solution. The two main steps of case-based reasoning are: (i) the "retrieve" step, which consists in finding, within the available cases stored in the database, the one(s) considered as most similar with the problem to solve; and (ii) the "reuse" step, which consists in adapting the solutions of the retrieved cases to solve the new problem.

Other complementary steps can be considered, as in [3], which is a reference paper in the domain. For instance, there may be a "revision" step, in which the generated solution is evaluated and possibly repaired. There is generally a "retainment" step, in which the solved problem is incorporated into the case base.

Various case-based reasoning systems have been implemented since the early works in this domain [45] and several approaches can be noticed. One of these approaches, called knowledge-intensive case-based reasoning in the literature [1, 2], consists in taking into account additional knowledge in the reasoning process. Two kinds of knowledge can be outlined: domain knowledge and adaptation knowledge. Domain knowledge defines concepts concerning the application domain, used in the case-based reasoning process, such as the domain vocabulary. Adaptation knowledge provides information useful for the evaluation of similar cases and for the adaptation of solutions, in for the "retrieve" and "reuse" steps.

Case-based reasoning has been considered as an alternative to rule-based systems, since it somehow reduces the knowledge acquisition process [29], since generally speaking cases are easier to obtain than rules. However this advantage is less relevant for knowledge-intensive case-based reasoning, where domain and/or adaptation knowledge has to be obtained and formalized. The work presented in [14], for instance, supports this point of view and proposes a formalization of domain and an adaptation knowledge using semantic web technologies.

Some works have proposed to combine case-based and knowledge-based reasoning methods. The main arguments found in literature for combining both approaches are: (i) achieving speedup learning and (ii) providing explanatory support to the case-based processes. For instance, in the CASEY system [30], the case-based method is used as first step, while the knowledge-based method (generally complex and slow), is used only in case of failure or the first step. Instead, in the BOLERO system [34], the case-based part provides meta-knowledge which is used to reduce the searching space used by rule-based part. The combination of the two approaches is also used in the CREEK system [2]mainly for providing explanatory support to the case-based processes.

The research community continues to be very active and various application domains are explored (see recent studies such as [28, 36]).

In the present framework, the cases are not raw data but rules, since they express already consolidated knowledge obtained by synthesis and validation of experimental data. Therefore, the case- and knowledge-based parts are fused, which is not considered in the above works. Moreover, usually the domain and adaptation knowledge used in a case-based reasoning system are acquired through knowledge elicitation in the classic way for knowledge-based systems. Another option would be to also learn that knowledge from the cases. This line of work is adopted in this chapter, since the reconciliation of the set of "cases" (the rules), which is part of the adaptation knowledge used in the prediction strategy, is learnt.

2.2. Data Reconciliation

Data reconciliation is one of the main problems encountered when different sources have to be integrated. It consists in deciding whether different data descriptions refer to the same real world entity (e.g. the same person or the same publication). The problem of data reconciliation has been known for more than six decades. It has been introduced by [39] and first formalized by [20]. Many approaches have been proposed in relational databases where the problem is referred as *record linkage*, *data cleaning* or *entity resolution* (see [18] for a survey).

The traditional approaches of data reconciliation consider the data description as structured in several attributes. To decide on the reconciliation or on the non-reconciliation of data, some of these approches use probabilistic models [21,58], such as Bayesian network or SVM. However, these probabilistic models need to be trained on labeled data. This training step can be very time-consuming what is not desirable in online applications. Indeed, in such online contexts, labeled data can not be acquired and runtime constraints are very strong. Alternative approaches have been proposed like [15] where the similarity measures are used to compute similarity scores between attribute values which are then gathered in a linear combination, like a weighted average. Although these approaches do not need training step, they however need to learn (starting from labeled data) some parameters as for example weights to be associated to similarity scores of the different attributes.

In Semantic Web, the problem of data reconciliation that is refer to as *data linking* has recently received a great attention. Many approaches have been proposed in order to link RDF data, in particular in the setting of Linked Open Data cloud (LOD)[1]. These approaches have been classified in different ways (see [22] for a survey). First, an approach can be considered as supervised or unsupervised. In some supervised approaches a user is needed to control the process of linking. For instance, in [6], the tool D-Dupe allows a user to choose similarity measures and to validate candidate links via a user interface. Other supervised approaches exploit training data in order to "learn" how to match instances [27, 41]. Unsupervised approaches usually use knowledge that is either declared in an ontology [25,47] or given by a domain expert [40,56]. Second, an approach can be considered as local (instance-based) or global (graph-based). In local approaches [27,40–42,56], each pair of instances are explored independently while in global approaches [25,47], discovered links affect the discovery of other links. Moreover, approaches use logical rules to discover high-quality links while other approaches are numerical and compute similarity scores between pairs of instances. Additionally, there exist approaches that use external resources, such as WordNet[2], to improve the data linking [47,49].

Among the existing approaches of data reconciliation, some [42,46,47,56] take into account knowledge translating the logical semantics of the ontology axioms, in particular (inverse) functionality of properties and keys (see [4,44,52] for examples of approaches that discover them automatically from data). Although several studies have considered the use of domain knowledge in classification, these concern fields where large amounts of data have to be treated, such as the semantic web [51] or image classification [35], with different issues, in particular the exploitation of topological knowledge [50] or of resource annotations [19]. In [50], the authors developed a query routing approach in a P2P network that is based on an aggregated view of the neighbors' content. This aggregated view is computed using a topological clustering algorithm. This is very different from the situation encountered in this chapter, characterized by "poor data", i.e., relatively few rules are available and have to be exploited in the best way.

[1] http://lod-cloud.net/

[2] http://wordnet.princeton.edu/

In [46, 47] a knowledge-based data reconciliation has been developed. The approach consists in a combination of two methods: the first method, called L2R, is logical and the second one, called N2R, is numerical. The two methods exploit knowledge that are declared in the schema that is expressed in OWL2. The exploited knowledge consists of a set of disjunction axioms between classes (e.g. Museum is disjoint with Painting), a set of (inverse) functional properties (e.g. a museum is located in at most one city) and the Unique Name Assumption. The logical method L2R is based on a set of logical inference rules that are generated from the translation of the logical semantics of the schema knowledge. The numerical method N2R is based on the computation of a similarity score for each part of data by using the values of the common properties. In the similarity computation, the functionality of the properties is exploited in a numerical way such that the similarity scores of the properties that are functional have a strong impact on the data pair similarity score. In both L2R and N2R a transitive closure is computed in order to obtain a partition of the set of reconciled data. Thus, each partition contains the set of data that refer to the same real world entity and two different partitions contain two data sets that refer to two different real world entities.

In the present work, two rules expressing the same (or very similar) experimental conditions (e.g. kind of water, cooking time) and results, may be considered as belonging to the same experimental tendency. Identifying domain rules that refer to the same experimental tendency is thus very close to the data reconciliation problem, and reconciliation techniques may prove to be suitable to this purpose. However, data reconciliation methods have to be adapted, since the reconciliation is no more applied on classical relational data, but on domain knowledge composed of causality rules enriched by ontological knowledge.

2.3. Previous Approaches in the Application Domain

Previous systems have been proposed in food science in order to help prediction. In particular, several works have dealt with the problem of food security, and therefore have proposed tools for risk assessment, such as in predictive microbiology, to prevent microbiological risk in food products. Such systems [5, 23, 43] combine a database with mathematical models which, applied to the data, allow one to deduce complementary information.

In the field of cereal transformation, we can also note the "Virtual Grain" project (see [37]), which gathers heterogeneous information concerning the properties of cereal grains and their agronomical conditions in a database in order to identify potential relationships between properties that are usually studied separately: morphological, biochemical, histological, mechanical and technological properties. The database is connected to statistical and numerical computing tools, in particular a wheat grain cartography tool developed using matlab. Based on wheat grain properties and information on the components distribution, it proposes a representation of components content in each tissue. However, the objectives of the Virtual Grain project are related to the explanation of the grain behavior during fractioning. Contrary to our concern, they do not include considerations on food products, industry process analysis and final quality prediction.

In the close domain of breadmaking technology, the "Bread Advisor" tool has been a pioneer as a knowledge software for the baking industries (see e.g. [60]). This tool, exclusively based on expert knowledge stored in a database, provides three kinds of information:

text information about the processing methods, list of possible faults and their causes, and generic messages about the effects of process changes. More recently, [38] have proposed a qualitative model of French breadmaking. However in both approaches knowledge from the scientific literature and dynamic prediction are not proposed.

The approach we propose has several original advantages from the application point of view. Firstly, concerning its objectives, it is not only a risk assessment system, but allows evaluating both defaults and qualities of food products. Secondly, although mostly tested on the durum wheat chain, it is not dedicated to a specific risk or chain, as it is not based on predetermined mathematical models, and therefore can be generalized to other application fields.

Finally, our approach for prediction is not based on predetermined domain models. Indeed, such models are often unavailable. Thus an important contribution of our work is to provide prediction solutions in case of lack of models.

Complementary considerations applied to the breadmaking food chain are studied in [7,12,54]. They deal with a different concern, namely multi-criteria decision making using reverse engineering, based on argumentation theory.

3. Basic Materials

In this section, we briefly recall essential elements regarding ontology and domain rule definition.

3.1. The Domain Ontology

The ontology O is defined as a tuple $O = \{C, \mathcal{R}\}$ where C is a set of concepts and $\mathcal{R}$ is a set of relations.

3.1.1. Ontology concepts

Each concept c is associated with a definition domain by the def function. This definition domain can be:

- *numeric*, i.e. $def(c)$ is a closed interval $[min_c, max_c]$;
- *'flat' (non hierarchized) symbolic*, i.e. $def(c)$ is an unordered set of constants, such as a set of bibliographic references;
- *hierarchized symbolic*, i.e. $def(c)$ is a set of partially ordered constants, themselves are concepts belonging to C.

In the sequel, we will refer to elements of concept domain definition by *values*.

3.1.2. Ontology relations

The set of relations $\mathcal{R}$ is composed of:

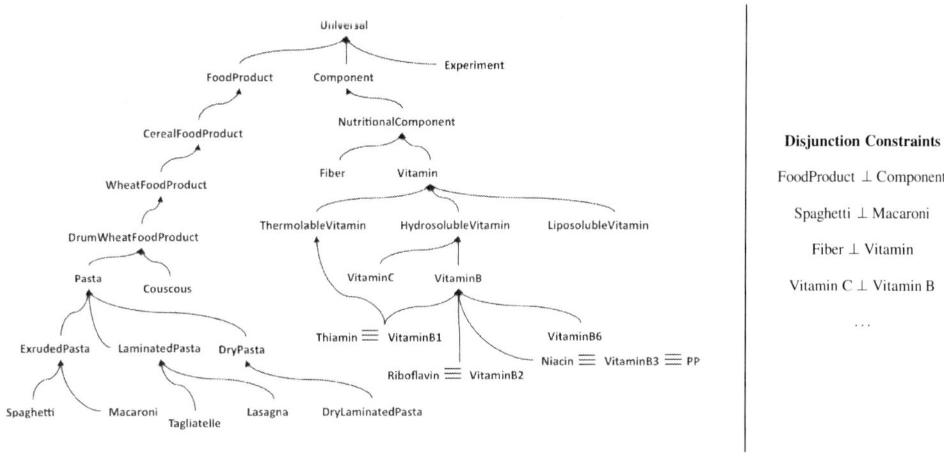

Figure 1. A part of the domain ontology

- the *subsumption* or 'kind of' relation denoted by $\preceq$, which defines a partial order over $\mathcal{C}$. Given $c \in \mathcal{C}$, we denote as $\mathcal{C}_c$ the set of sub-concepts of c, such that: $\mathcal{C}_c = \{c' \in \mathcal{C} | c' \preceq c\}$. When c is defined by hierarchized symbolic definition domain, we have $def(c) = \mathcal{C}_c$.

- the *equivalence* relation, denoted by $\equiv$, expressing a synonymy between concepts of the ontology.

- the *disjunction* relation between concepts, denoted by $\perp$. Given two concepts c and $c' \in \mathcal{C}$, $c \perp c' \Rightarrow (def(c) \cap def(c')) = \emptyset$. We note that the disjunction relation respects the subsumption relation. This means that if two general concepts c and c' are declared as disjoint then all the concepts that are more specific than c and c', respectively, are pairwise disjoint.

Figure 1 gives a small part of the set of concepts $\mathcal{C}$, partially ordered by the subsomption relation (pictured by '$\rightarrow$'). Examples of disjunctions are given apart for readability reasons. Note that the considered ontologies are not restricted to trees, they are general graphs. This is an important feature of our work with respect to previous approaches, such as [61], where only trees are considered.

3.1.3. Least common subsumer

Given two concepts c_1 and c_2, we denote as $lcs(c_1, c_2)$ their least common subsumer, that is $lcs(c_1, c_2) = \{c \in \mathcal{C} | c_i \preceq c$, and $((\exists c'$ s.t. $c_i \preceq c') \Rightarrow (c \preceq c'))$, $i \in \{1, 2\}\}$.

For example, in the ontology of Figure 1, the *lcs* of the concepts *LiposolubleVitamin* and *VitaminB* is the concept *Vitamin*. As commonly done, we consider a *Universal* concept subsuming all the other concepts of the ontology, to ensure that such a *lcs* always exists.

3.1.4. Relationship between ontology concepts and experimental variables

We consider a set of experimental descriptions containing K variables. Each variable X_k, $k = 1,\ldots,K$, is associated with a concept $c \in C$ of the ontology O. Each variable can be instantiated by a value that belongs to the definition domain of concept c.

3.1.5. Variable discriminance

For each variable X_k, a discriminance score, denoted by λ_k, is declared. It is a real value in the interval $[0;1]$. It is obtained through an iterative approach performed with domain experts, as briefly explained in section 4.3 and in more detail in [55].

3.2. Domain Rules

Each domain rule expresses the relation cause-to-effect between a set of parameters (e.g. *KindOfWater*, *CookingTime*) of a unit operation belonging to the transformation process, and the variation of a given product property (e.g. variation of *vitamin content*). The set of domain rules is denoted by $\mathcal{R}$.

Definition 1 (Causality rule). *A causality rule $R \in \mathcal{R}$ is defined by a pair (H,C) where:*

- $H = \{(X_1,v_1),\ldots,(X_h,v_h)\}$ *corresponds to the rule hypothesis. It is a conjunction of variable/value criteria describing a set of experimental conditions in the form* $(X_i = v_i)$. *The value v_i may take numeric or symbolic (flat or hierarchized) values.*
- $C = (X_c,v_c)$ *corresponds to the rule conclusion that is composed of a single variable/value effect describing the resulting impact on the considered property.*

It is interpreted by:

$$(X_1 = v_1) \wedge (X_2 = v_2) \wedge \ldots \wedge (X_h = v_h) \Rightarrow (X_c = v_c)$$

Table 1 shows a part of the set of rules of the considered case study. The experimental variables are given in the first line and the values of these variables are given in the other lines of the table. The last column represents the conclusion part of the rules (i.e., the variable X_c and its values). For example, the rule $R1$ given in the second line is interpreted as follows:

(CookingTime = 12) ∧ (KindOfWater = Tap Water) ∧ (Component = Riboflavin)) ⇒ (ConcentrationVariation = -53.3).

In the following, domain rules will be compared on the basis of the set of variables they have in common, called common description and defined as follows.

Definition 2 (Common description). *A common description between two causality rules R_1 and R_2, denoted by* CommonDesc(R_1, R_2), *consists of the set of variables appearing at the same time in the set of variables/values of R_1 and R_2. Let H_1 and H_2 be the hypotheses of the rules R_1 and R_2 respectively. Let C_1 and C_2 be their conclusions.*

$CommonDesc(R_1,R_2) = \{X \mid \exists v_1, v_2,\ [(X = v_1) \in H_1\ \ and (X = v_2) \in H_2]\ \ or\ [(X = v_1) = C_1\ and\ (X = v_2) = C_2]\}$.

We denote as $CommonDesc_H(R_1,R_2)$ the common description of R_1, R_2 reduced to their hypotheses, i.e. $CommonDesc_H(R_1,R_2) = \{X \mid \exists v_1, v_2,\ [(X = v_1) \in H_1\ and\ (X = v_2) \in H_2]\}$.

Table 1. ***A set of causality rules***

Id.	CookingTime (min)	KindOfWater	Component		ConcentrationVariation (%)
R1:	12	Tap water	Riboflavin	⇒	-53.3
R2:	12	Tap water	Niacin	⇒	-45.6
R3:	12	Tap Water	Vitamin B	⇒	-45.6
R4:	24	Tap Water	Fiber	⇒	-45.6
R5:	13	Water	Vitamin B6	⇒	-46
R6:	10	Deionized water	Thiamin	⇒	-52.9
R7:	10	Deionized water	Vitamin B1	⇒	-51.8
R8:	15	Distilled water	Vitamin B2	⇒	-45.5

4. Domain Rule Reconciliation

This section presents a method for domain rule reconciliation that partitions the set of rules by using techniques adapted from data reconciliation. We first give the general principle of rule partitioning approach. Then, we present how rules are filtered, compared and finally grouped into several groups of similar rules.

4.1. Principle

The general principle behind rule reconciliation consists in determining which rules can be considered as similar, thus providing a partition of the given set of rules. The partition is computed according to a specific similarity measure. Some rules are considered as similar if their similarity value is above a certain threshold, otherwise they are called dissimilar. Similar rules are positioned in the same reconciliation group (partition subset).

Obviously, the key problem is related to the definition of an appropriate similarity measure which must take into account several features. In the literature, classic similarity measures are used for basic parameter values as in [10]. The choice of these similarity measures is done according to the features of the values, e.g. symbolic/numeric, length of values, and so on. It is also possible to consider a semantic similarity measure that exploits the hierarchy distance of the values in a given domain ontology (see Wu and Palmer measure [59]). In a complementary approach, additional ontological knowledge, as synonymy relations between concepts, can be exploited. Finally, it is possible to compute the importance of the variables which appear in the rules, or take into account the declared knowledge, as functional dependencies or keys that can be discovered automatically as proposed in [52, 57].

4.2. Ontology-based Filtering Step

A pre-processing step relies on knowledge declared in the ontology. In this step, we exploit the semantics of the disjunction relations between concepts. For example, the concepts *ExtrudedPasta* and *LaminatedPasta* of Figure 1 are declared as disjoint. In this step we claim that two rules containing disjoint variable values cannot belong to the same group, and they are defined as "disjoint".

Definition 3 (Disjoint causality rules). *Two rules R_1 and R_2 are said to be disjoint if and only if:*

- *either* $CommonDesc(R_1, R_2) = \emptyset$*;*
- *or* $\exists X \in CommonDesc(R_1, R_2)$ *such that, given* v_1*,* v_2 *the respective values of* X *in* R_1 *and* R_2*, the disjointness constraint* $v_1 \bot v_2$ *can be inferred from the ontology.*

For example, in Table 1, the rule R4 is disjoint from all the other rules because - in the ontology - the value Fiber of the Component variable is declared as disjoint from the Vitamin(see Fig.1). All the other values taken by the *Component* variable are sub-concepts of *Vitamin* in the ontology, and therefore disjoint from *Fiber* since the disjointness constraint is propagated to all the sub-concepts of *Vitamin*.

4.3. Variable Relevance in Rule Similarity Computation

In order to associate each variable X_k with a score λ_k reflecting its discriminance power, there are two possible strategies: (i) ask a domain expert to specify such knowledge or (ii) obtain it automatically by using a learning method. In this work, we adopted both approaches in a complementary way. Thus the computation of the score λ_k is adapted from the N2R method proposed in [47], but it is determined using both declared expert knowledge and supervised learning, in the following way:

- Functional dependencies are declared knowledge, obtained through a collaboration with domain experts. For instance, in the considered application, a subset of the variables describing the experimental conditions has been identified by the experts as determining the *ConcentrationVariation* variable, i.e. the one to be predicted. These variables are: *Temperature*, *SaltPercentage*, *CookingTime*, *KindOfWater*, *Component*, *AdditionOfIngredients*. The *FoodProduct*, *hasMethod*, and *ValueBefore* variables do not belong to this list. We propose to take into account this knowledge by associating a low discriminance score with the latter variables (the chosen weight is 0 in the practical evaluation of section 6., i.e. the variables are ignored).
- In order to distinguish between the variable importance among those variables that belong to functional dependencies an iterative method in interaction with experts is applied (see [55] for details).

4.4. Similarity Measures

In order to compute the similarity of two causality rules, a prerequisite is to measure the similarity of the pairs of values, in an arbitrary order, belonging to their common descriptions.

4.4.1. Similarity measures between values

To measure the similarity between values, we use existing similarity measures. A similarity measure is usually defined as follows.

Definition 4 (Similarity measure). *Given a universe of discourse $\mathcal{U}$, a similarity measure Sim is a function $\mathcal{U} \times \mathcal{U} \rightarrow \mathbb{R}$ that satisfies the following properties:*

- *positivity:* $\forall v_1, v_2 \in \mathcal{U}, Sim(v_1, v_2) \geq 0$
- *symmetry:* $\forall v_1, v_2 \in \mathcal{U}, Sim(v_1, v_2) = Sim(v_2, v_1)$
- *maximality:* $\forall v_1, v_2 \in \mathcal{U}, Sim(v_1, v_1) \geq Sim(v_1, v_2)$.

Other properties may be required, like normalisation that will be assumed here, which imposes the *Sim* function to take values in the interval $[0; 1]$.

In [32] a survey and a deep analysis of similarity measures is presented, while in [10] experimental approaches are summarized. Since there is no universal measure which can be qualified as the most efficient for every kind of value, the choice of a suitable similarity measure depends on the features of the considered basic values, as for example, symbolic/numeric values or length of the values.

We present in detail the choices we made in the application section (see section 6. of this chapter). Here, we outline that three cases may be distinguished, depending on the kind of the definition domain of the considered variables.

Hierarchized symbolic variables: a semantic similarity measure that exploits the hierarchy distance of the property values in the domain ontology may be used, as Wu and Palmer [59] and Lin [33]. A framework for similarity measures in ontologies is proposed by [17]. We retained the Wu and Palmer measure, which is more intuitive and easy to implement. Its principle is based on the length of the path between two concepts in the hierarchy. Its definition is as follows:

$$Sim(v_1, v_2) = \frac{2 * depth(v)}{depth_v(v_1) + depth_v(v_2)},$$

where v is the least common subsumer (lcs) of the values v_1 and v_2, $depth(v)$ is the length of the path (number of arcs) between v and the root node of the hierarchy, and $depth_v(v_i)$ is the length of the path between v_i and the root node through the node v.

For example, in the ontology of Figure 1, the semantic similarity of the values $(Spaghetti, Lasagna)$ by using the Wu and Palmer measure is computed as follows: $Sim(Spaghetti, Lasagna) = \frac{2*5}{7+7} = 0.71$.

Moreover we pose: $v_1 \equiv v_2 \Rightarrow Sim(v_1, v_2) = 1$.

For instance, $Sim(Thiamin, VitaminB1) = 1$ by hypothesis, since we have $Thiamin \equiv VitaminB1$.

'Flat' (non hierarchized) symbolic variables: numerous similarity measures have been proposed in the literature. One of the most exhaustive inventory is given in [9]. We use the Braun & Banquet similarity measure, which is defined as follows.

Let V_1 be the set of elements (in our case, words) composing v_1 and V_2 the set of elements composing v_2:

$$Sim_{Braun\&Banquet}(v_1, v_2) = \frac{|V_1 \cap V_2|}{max(|V_1|, |V_2|)}.$$

This measure is related to the well-known Jaccard coefficient: $Sim_{Jaccard}(v_1, v_2) = \frac{|V_1 \cap V_2|}{|V_1 \cup V_2|}$. Both Braun & Banquet and Jaccard similarity measures take the value 0 when V_1 and V_2 have no common elements, and the value 1 when V_1 and V_2 are equal. However in intermediate cases $Sim_{Braun\&Banquet} \geq Sim_{Jaccard}$. For instance, for $v_1 =$ *"deionized water"* and $v_2 =$ *"distilled water"*, we have $Sim_{Braun\&Banquet}(v_1, v_2) = \frac{1}{2}$ and $Sim_{Jaccard}(v_1, v_2) = \frac{1}{3}$. The mismatch between *"deionized"* and *"distilled"* is sanctioned twice in the Jaccard measure, since its denominator indicates the number of distinct words ("deionized", "distilled", "water"), whereas in the Braun & Banquet denominator the mismatch is only counted once, which we privileged here.

Numeric variables: a synthesis on similarity measures is presented in [31]. Let X be a numeric variable, v_1 and v_2 two values of $Range(X)$. The similarity measure we use is defined in a classical way, as the complement to 1 of a normalised distance:

$$Sim(v_1, v_2) = 1 - \frac{|v_2 - v_1|}{|Range(X)|}.$$

4.4.2. Similarity measures between two causality rules

In this work, the similarity between two rules is computed as a weighted average of the similarity scores of the values taken by their common variables. We denote as $Similarity(R_1, R_2)$ the real function which measures the similarity between two causality rules R_1 and R_2. It is computed as follows:

Definition 5 (Similarity of two causality rules).

$$Similarity(R_1, R_2) = \sum_k \lambda_k * Sim(v_{k_1}, v_{k_2}),$$

where $k \in [1 \; ; \; |CommonDesc(R_1, R_2)| \;]$*,* λ_k *is the discriminance power of the variable* $X_k \in CommonDesc(R_1, R_2)$*, and* v_{k_1}, v_{k_2} *are the values taken by* X_k *in* R_1 *and* R_2 *respectively.*

Illustrative example of two causality rules similarity. Let us consider $R1$ and $R5$ of Table 1, two rules which are not disjoint. To compute their similarity score we perform the following steps.

1. compute the common description : $CommonDesc(R1, R5) = \{CookingTime, KindOfWater, Component, ConcentrationVariation\}$;
2. determine the discriminance power of the considered variables. According to the assumptions of the sub-section 4.3. all the considered variables have a high discriminance power. Therefore the scores λ_k equal to $1/4$;
3. compute the elementary similarity scores:

   $S_1 = Sim(12, 13) = 0.93$, by using the similarity measure between numeric values.
   $S_2 = Sim(Riboflavin, VitaminB6) = \frac{2*5}{6+6} = 0.83$, by using the Wu & Palmer measure.
   $S_3 = Sim(-53.3, -46) = 0.84$, by using the similarity measure between numeric values.

$S_4 = Sim(\text{"Tap Water"}, \text{"Water"}) = 0.5$ by using the Braun & Banquet similarity measure.

$Similarity(R1, R5) = \frac{1}{4} * (S_1 + S_2 + S_3 + S_4) = 0.77.$

4.5. Partitioning the Set of Rules into Reconciliation Groups

To decide which rules must be reconciled, we use a similarity threshold and assign to the same group each pair of rules having a similarity score beyond this given threshold. Rules having a similarity score beyond the threshold express a common experimental tendency.

We denote as $Reconcile(R_1, R_2)$ the predicate which expresses that the causality rules R_1 and R_2 express the same experimental tendency and must be reconciled. Let th be the similarity threshold for rule reconciliation. The reconciliation decision, for each pair of causality rules (R_1, R_2), is expressed as follows:
If $Similarity(R_1, R_2) \geq th$, then $Reconcile(R_1, R_2)$.
For example, the rules $R1$ and $R5$ (see their similarity score in the above section) will be reconciled if we consider a threshold th lower than 0.77.

We note that the threshold is fixed experimentally by exploiting the size and the homogeneity nature of the rule set (see Section 6.). Finally, to obtain disjoint groups of rules, we compute a transitive closure for the set of reconciliation decisions previously computed. This choice, developed in [47], could be relaxed in the future by considering a fuzzy framework with overlapping groups. The transitive closure can be computed by using the following rule (TC):

$$\forall R_1, R_2, R_3,\ Reconcile(R_1, R_2) \wedge Reconcile(R_2, R_3) \Rightarrow Reconcile(R_1, R_3).$$

Algorithm 1 presents the reconciliation method. We can note that:

1. the result does not depend on the order of comparisons between rules;
2. the decision of reconciliation is based on pairwise comparisons between rules;
3. the algorithm does not guarantee a minimum number of rules per group, i.e. it is not excluded to end up with:
   - a rule group containing only a single rule, or
   - a single group for all the rules.

In Algorithm 1, we used several additional functions that can be defined as follows:

- the function $Disjoint(R_i, R_j)$ returns a boolean value indicating whether the rules R_i and R_j are disjoint or not, as defined in Definition 3;
- the function $Similarity_{\Lambda, \mathcal{S}}(R_i, R_j)$ computes the similarity between the rules R_i and R_j by using the discriminance scores Λ and the similarity measures $\mathcal{S}$, as defined in Definition 5;
- the function $TransitiveClosure(G)$ computes the transitive closure of G by applying the transitivity rule (TC) on the set of computed groups.

Algorithm 1: RECONCILIATION ALGORITHM
Input:

- $\mathcal{R}$: a set of causality rules, defined using the list of variables $\mathcal{X}$
- Λ: a list of discriminance scores associated with the list of variables $\mathcal{X}$
- $\mathcal{S}$: a list of similarity measures specified for the list of variables $\mathcal{X}$
- th: a threshold

Output: G: a set of groups of reconciled rules

$G \leftarrow \emptyset$
$PairSim \leftarrow 0$ // similarity between two rules
foreach $(R_i, R_j) \in \mathcal{R} \times \mathcal{R}$, with $(i \neq j)$
    if $(Disjoint(R_i, R_j) =$ false$)$
        $PairSim \leftarrow Similarity_{\Lambda,\mathcal{S}}(R_i, R_j)$
        if $(PairSim \geq th)$
            if $(\exists g_i \in G,\ R_i \in g_i)$
                $g_i \leftarrow g_i \cup \{R_j\}$
            else if $(\exists g_j \in G,\ R_j \in g_j)$
                $g_j \leftarrow g \cup \{R_i\}$
            else
                $g \leftarrow \{R_i, R_j\}$
                $G \leftarrow G \cup \{g\}$
    $G \leftarrow TransitiveClosure(G)$

5. Predicting New Rules

This section presents the predictive step of the proposed method.

5.1. Principle

We consider G and R. G is the set G of reconciliation groups computed as presented in the previous section, and R is a new rule whose conclusion value is still unknown, and which has a set of variable/value criteria corresponding to its hypothesis. The new rule R is called an "unknown output rule", denoted by "UO-rule", and it is in the following form:

$R = (H, C)$ with $H = \{(X_1 = v_1), \ldots, (X_H = v_H)\}$ and $C = (X_c = x)$, where x is unknown.

The objective of this step is to predict the conclusion value x of the UO-rule R, that is, the expected variable/value effect, by comparing it to the existing known rules.

The method consists in selecting the closest reconciliation group and then use it to predict the conclusion value x. To perform this stage, R is compared to representative member(s) – called the "kernel" – of each reconciliation group. R is then assumed to be part of the group g whose kernel contains the rule the most similar to R. The value of the conclusion of R is predicted as a combination of the conclusion values of the rules of g.

5.2. Kernel Rule(s) of a Group

Both for combinatory and for semantic reasons, a *kernel* is computed for each reconciliation group, which corresponds to the set of rules that are most representative of the group. From a combinatory point of view, comparing the UO-rule R to the kernel of each group is of course simpler than comparing it to the whole set of rules. From a semantic point of view, the kernel can be seen as a characteristic sample of a reconciliation group. Aggregating multiple knowledge sources by retaining a well-chosen piece of them is a common feature in knowledge fusion [16]. An aggregation framework close to our approach, in a totally different domain, is proposed in [11].

The kernel of a group g is defined as follows.

Definition 6 (Kernel). *Let g be a group of rules and $R_i \in g$. The representativity of R_i is measured by:*

$$Repr(R_i) = \Sigma_{j=1\ (j \neq i)}^{j=|g|} Similarity(R_i, R_j).$$

The kernel of g is then defined by:

$$Kernel(g) = \{r \in g \mid Repr(r) = \max_{R_i \in g} Repr(R_i)\}.$$

The most representative rule is thus chosen (or several ones in case of *ex aequo*). Note that the set $Kernel(g)$ is often reduced to a unique rule.

5.3. Allocation of R to the Closest Reconciliation Group

To compare R with the rule(s) of $Kernel(g)$, a similarity score is computed as in the Definition 5. The difference is that only the variable/value criteria appearing in the rule hypotheses are taken into account, since the conclusion value is unknown for R.

Definition 7 (Similarity between a UO-rule and a rule). *Let R be a UO-rule and R' a rule.*

$$Similarity(R, R') = \sum_k \lambda_k * Sim(v_k, v'_k),$$

where $k \in [1\ ;\ |CommonDesc_H(R, R')|\ ]$, λ_k is the discriminance power of the variable $X_k \in CommonDesc_H(R_1, R_2)$, and v_k, v'_k are the values taken by X_k in R and R' respectively.

R is then allocated to the group g whose kernel contains the rule the most similar to R.

Definition 8 (Group allocated to a UO-rule). *Let G be a set of reconciliation groups and R a UO-rule. R is allocated to a group $g \in G$, which satisfies the following condition:*

$$\exists R' \in Kernel(g), Similarity(R, R') = \max_{r \in Kernel(g_i), i \in [1;|G|]} Similarity(R, r).$$

5.4. Prediction Method

To predict the conclusion value x of R, two cases are distinguished, according to the nature – symbolic or numeric – of the conclusion variable X_c.

Symbolic case. It is the case where the definition domain of the considered variable is hierarchized or flat symbolic. The conclusion variable X_c may take several values in the reconciled rules of group g. We make the assumption that the value x to be predicted figures among those already appearing in the group g, i.e. that X_c does not take a new value in the conclusion of R.

For each distinct value v_i taken by X_c in the group g, a confidence degree $conf_{v_i}$ (a real value in [0;1]) is computed. It can be interpreted as the confidence in the prediction that "the value taken by X_c in R is v_i". It is defined as the ratio between the sum of similarity scores between R and the rules where X_c takes the value v_i, and the total sum of similarity scores between R and the rules of g, that is:

$$conf_{v_i} = \frac{\sum_{\{r \in g | X_c = v_i\}} Similarity(R,r)}{\sum_{\{r \in g\}} Similarity(R,r)}.$$

The value with the highest confidence provides the prediction of the conclusion value x:

$$(\tilde{x} = v) \ such\ that\ conf_v = \max_i\ conf_{v_i},$$

where $\tilde{x}$ denotes the value predicted for x.

Numeric case. The prediction of the value x taken by X_c in R is computed as a weighted mean of the values taken by X_c in all the rules of the group g. The weight associated with each rule r of g is the similarity score between R and r. Let v_r be the value taken by X_c in r, the value x is predicted by:

$$\tilde{x} = \frac{\sum_{r \in g} (Similarity(R,r) \times v_r)}{\sum_{r \in g} Similarity(R,r)}.$$

A confidence degree $conf$ is attached to the prediction. This confidence degree is the weighted mean of the similarity scores between R and the rules of g:

$$conf = \frac{\sum_{r \in g} (Similarity(R,r))^2}{\sum_{r \in g} Similarity(R,r)}.$$

Algorithm 2 presents the prediction method. For the sake of simplicity, only the numeric case is considered in the following algorithm.

Algorithm 2: PREDICTION ALGORITHM

Input:

- $G = \{g_1, \ldots, g_{|G|}\}$: a set of groups of reconciled rules
- $R = (H, C)$: a UO-rule with $H = \{(X_1 = v_1), \ldots, (X_H = v_H)\}$ and $C = (X_c = x)$, where the value x is unknown

Output: $(\tilde{x}, conf)$: with $\tilde{x}$ a prediction of the value x and $conf$ a confidence degree for the prediction $\tilde{x}$

$S \leftarrow 0$ // the best similarity between kernel rules and R
$g \leftarrow null$ // the selected group
foreach $\{g_i \in G\}$
  foreach $r \in Kernel(g_i)$
    if $(S < Similarity(R, r))$
      $S \leftarrow Similarity(R, r)$
      $g \leftarrow g_i$
$\tilde{x}\_num \leftarrow 0$ // numerator of $\tilde{x}$: weighted sum of the conclusion values of the rules of g
$conf\_num \leftarrow 0$ // numerator of $conf$: sum of the squares of similarities between R and the rules of g
$denom \leftarrow 0$ // denominator of $\tilde{x}$ and $conf$: sum of similarities between R and the rules of g
foreach $\{r = (h, c) \in g\}$, with $c = (X_c, v_c)$
  $s \leftarrow Similarity(R, r)$
  $\tilde{x}\_num \leftarrow \tilde{x}\_num + s \times v_c$
  $conf\_num \leftarrow conf\_num + s^2$
  $denom \leftarrow denom + s$
$\tilde{x} \leftarrow \tilde{x}\_num \div denom$
$conf \leftarrow (conf\_num \div denom)$

In Algorithm 2, we used several additional functions that can be defined as follows:

- the function $Kernel(g_i)$ selects the set of kernel rules of g_i, according to Definition 6;
- the function $Similarity(R, r)$ computes the similarity between the UO-rule R and the rule r, according to Definition 7;
- the function $Size(g_i)$ returns the number of rules in the group g_i.

A discussion as well as complexity results are provided in [48].

6. Application in Food Science

In this section, the application domain and the adopted evaluation protocol are described. Then the results are presented, providing a qualitative and quantitative evaluation.

6.1. Context of the Case Study

The efforts in cereal food design during the last 20 years resulted in an explosion of scientific papers which can hardly be used in practice because they have not been completely integrated into a corpus of knowledge. At the same time, cereal and pasta industry has evolved from traditional companies to a dynamic industry geared to follow consumer trends: healthy, safe, easy to prepare, pleasant to eat [13]. To meet such criteria, current industry requires knowledge from its own know-how as well as from different disciplines. Moreover, it has to extract the key information from the available knowledge in order to support decision.

Knowledge on the domain of transformation and properties of cereal-based food is available through the international literature of the domain concerning the impact of the transformation process on food properties. More precisely, the following elements are described:

- the "technical" information concern the unit operations which are involved in transformation from the wheat grains to the ready-to-use food (e.g. grinding, storage, drying, baking, etc.);
- the "property" information define the criteria which are used to represent the properties of products, according to three aspects: organoleptic, nutritional and safety properties (e.g. colors, vitamins contents, pesticides contents, etc.);
- the "result" information provide the impact of a unit operation on a property (i.e. what happens at the "intersection" between an element of the technical information and an element of the property information).

For each unit operation composing the transformation process, and for each family of product properties, this knowledge has been expressed as causality rules [53]. Approximately 600 rules are available in the application.

6.2. Evaluation Protocol

The evaluation was made in collaboration with food science experts. The proposed predictive method was compared with a classic decision tree prediction approach. In this chapter, we use C4.5 recursive dichotomous partitioning [8] in the R software [26] rpart implementation.

The methodology used to evaluate the proposed method followed six steps:

1. definition of requirements on the quantity and form of the test queries (i.e. UO-rules), so that the results are significant;
2. definition of requirements on the rule base, so that the results are interpretable;
3. definition of a set of test queries that satisfy the previous requirements;
4. definition of the parameters used by the methods;
5. execution of the set of test queries;
6. analysis of the results.

In the following we describe the procedure step by step.

6.2.1. Significance Conditions on the Test Queries

The significance conditions put on the test queries aimed at ensuring that the set of test queries is as heterogeneous and representative as possible. They are of six kinds:

- the set of test queries should cover at least thirty percent of the rule base entries (value chosen because a 70-30 ratio is often used in data exploration);
- the set of test queries should cover all branches of the ontology;
- the set of test queries should include specific rules (i.e. without missing data), as well as general rules (with missing data);
- the set of test queries should include rules that have an exact answer in the queried rule base, as well as rules that are absent from the queried rule base. The latter should cover cases where there is a close answer in the queried rule base, or no close answer in the queried rule base;
- the values used in the test queries should be taken from an external rule base, whose conclusion values are known, so that the results can be objectively evaluated;
- several thresholds should be tested.

6.2.2. Interpretability Conditions on the Rule Base

The interpretability conditions put on the rule base were the following: several (at least two) different rule bases should be used, in order to introduce a variability in the rule base content. Considering its impact on the method results will allow testing the robustness of the method.

6.2.3. The Set of Test Queries

The test queries are defined as UO-rule hypotheses. The objective is to predict their conclusion values. Considering the above conditions, the test queries presented in Table 2 were defined.

6.2.4. Method Parameters

The predictive method presented in this chapter was used with two different thresholds for partitioning, respectively 0.8 and 0.9.

The implementation used for decision trees is the R software with the *rpart* package for CART trees. The parameters of the rpart algorithm are: cross validation = 100, minimum instances per leaf = 6 (default value).

6.2.5. Experimental Results

Each of these methods was executed on two different rule bases. Rule base 1 contains 109 rules. Rule base 2 contains 117 rules. All of them concern the "cooking in water" unit operation and the "vitamin content" and "mineral content" properties. The execution of the

Table 2. Set of test queries

Q1	{(FoodProduct=Spaghetti), (Component=Riboflavin), (ValueBefore= 0.6), (Temperature= 100), (SaltPercentage= 0), (Time= 12), (KindOfWater=Tap water), (Riboflavin addition=0.65), (Drying cycle=LT), (Drying time=28), (Max drying temperature=39)}
Q2	{(FoodProduct=Spaghetti), (Component=Riboflavin), (ValueBefore=0.4), (Temperature=100), (SaltPercentage=0), (Time=24), (KindOfWater=Tap water), (Riboflavin addition=0.65), (Drying cycle=HT-B), (Drying time=14), (Max drying temperature=70)}
Q3	{(FoodProduct=Spaghetti), (Component=Riboflavin), (ValueBefore=1.63), (Temperature=100), (SaltPercentage=0), (Time=12), (KindOfWater=Tap water), (Riboflavin addition=1.95), (Drying cycle=HT-B), (Drying time=14), (Max drying temperature=70)}
Q4	{(FoodProduct=Spaghetti), (Component=Thiamin), (ValueBefore=1), (Temperature=100), (SaltPercentage=0), (Time=12), (KindOfWater=Tap water), (Thiamin addition=0.96), (Drying cycle=HT-A), (Drying time=13), (Max drying temperature=75)}
Q5	{(FoodProduct=Spaghetti), (Component=Niacin), (ValueBefore=5.7), (Temperature=100), (SaltPercentage=0), (Time=24), (KindOfWater=Tap water), (Niacin addition=2.24), (Drying cycle=LT), (Drying time=28), (Max drying temperature=39)}
Q6	{(FoodProduct=Spaghetti), (Component=Niacin), (ValueBefore=9.6), (Temperature=100), (SaltPercentage=0), (Time=12), (KindOfWater=Tap water), (Niacin addition=6.72), (Drying cycle=LT), (Drying time=28), (Max drying temperature=39)}
Q7	{(FoodProduct=Spaghetti), (Component=Niacin), (ValueBefore=9.6), (Temperature=100), (SaltPercentage=0), (Time=24), (KindOfWater=Tap water), (Niacin addition=6.72), (Drying cycle=HT-A), (Drying time=13), (Max drying temperature=75)}
Q8	{(FoodProduct=Macaroni), (Component=Thiamin), (ValueBefore=11.4), (Temperature=100), (Time=10), (KindOfWater=Deionized water)}
Q9	{(FoodProduct=Noodles), (Component=Folic acid), (ValueBefore=0.026), (Temperature=100), (SaltPercentage=0), (Time=14)}
Q10	{(FoodProduct=Macaroni), (Component=Vitamin B6), (ValueBefore=1.129), (Temperature=100), (SaltPercentage=0), (Time=14)}
Q11	{(FoodProduct=Pasta), (Component=Thiamin), (ValueBefore, 1.08), (Temperature, 100), (SaltPercentage, 0), (KindOfWater, Distilled deionized water)}
Q12	{(FoodProduct=Pasta), (Component=Riboflavin), (ValueBefore=0.43), (Temperature=100), (SaltPercentage=0), (KindOfWater=Distilled deionized water)}
Q13	{(FoodProduct=Pasta), (Component=Niacin), (ValueBefore=7.82), (Temperature=100), (SaltPercentage=0), (KindOfWater=Distilled deionized water)}
Q14	{(FoodProduct=Noodles), (Component=Phosphorous), (ValueBefore=201.2), (Temperature=100), (SaltPercentage=0), (Time=8), (KindOfWater=Tap water)}
Q15	{(FoodProduct=Noodles), (Component=Potassium), (ValueBefore=227.2), (Temperature=100), (SaltPercentage=0.5), (Time=8), (KindOfWater=Tap water)}
Q16	{(FoodProduct=Noodles), (Component=Calcium), (ValueBefore=27.2), (Temperature=100), (SaltPercentage=0), (Time=8), (KindOfWater=Tap water)}
Q17	{(FoodProduct=Noodles), (Component=Magnesium), (ValueBefore=56.6), (Temperature=100), (SaltPercentage=0.5), (Time=8), (KindOfWater=Tap water)}
Q18	{(FoodProduct=Noodles), (Component=Iron), (ValueBefore=3.4), (Temperature=100), (SaltPercentage=0), (Time=8), (KindOfWater=Tap water)}
Q19	{(FoodProduct=Noodles), (Component=Manganese), (ValueBefore=0.7), (Temperature=100), (SaltPercentage=0.5), (Time=8), (KindOfWater=Tap water)}
Q20	{(FoodProduct=Noodles), (Component=Zinc), (ValueBefore=1.6), (Temperature=100), (SaltPercentage=0), (Time=8), (KindOfWater =Tap water)}

set of test queries gave the results presented in Table 3 (a). The queries were executed both using the reconciliation method presented in this chapter, and using the decision tree (DT) method.

6.3. Result Analysis

In Table 3 (b), the error rates obtained with each method are computed. In Figures 2 (a) and (b), these error rates are presented as histograms, respectively for rule base 1 and rule base 2. The left columns represent the error rates obtained with the proposed method, for the two tested thresholds. The right columns represent the error rate obtained with the decision tree method.

These results clearly show a lower error rate obtained with the proposed predictive method, in all cases.

Table 3. (a) Execution results and (b) Average error rates, for the proposed method vs decision trees (DT)

(a)

Query	Rule base 1					Rule base 2					Expected
	Proposed method				DT	Proposed method				DT	
	Threshold 0.8		Threshold 0.9			Threshold 0.8		Threshold 0.9			
	pred.	conf.	pred.	conf.		pred.	conf.	pred.	conf.		
Q1	-57.16	0.72	-66.71	0.91	-57	-41.65	0.65	-43.1	0.82	-54	-53.3
Q2	-58.58	0.72	-60.05	0.89	-64	-41.77	0.65	-44	0.81	-49	-62.5
Q3	-57.32	0.73	-59.83	0.88	-56	-41.69	0.65	-43.26	0.81	-60	-50.9
Q4	-56.44	0.78	-64.3	0.88	-57	-42.56	0.69	-43.01	0.81	-49	-54
Q5	-47.98	0.79	-45.6	0.93	-63	-42.03	0.57	-58.14	0.81	-48	-42.1
Q6	-47.95	0.83	-53.1	0.93	-55	-42.48	0.57	-57.35	0.79	-67	-57.3
Q7	-48.23	0.85	-50.04	0.89	-61	-43.06	0.59	-58.18	0.75	-47	-59.4
Q8	-40.51	0.52	-43.93	0.94	-45	-42.72	0.62	-53.73	0.68	-35	-42.2
Q9	-23.46	0.81	-21	0.86	-37	-41.65	0.66	-21	0.87	-13	-21
Q10	-47	0.85	-47	0.85	-40	-41.7	0.68	-35	0.97	-12	-44
Q11	-48.99	0.72	-49.2	0.94	-46	-42.65	0.69	-52.56	0.77	-38	-49.3
Q12	-56.33	0.67	-37	0.83	-46	-42.53	0.7	-52.3	0.77	-37	-40.2
Q13	-47.85	0.78	-35	0.83	-46	-43.37	0.63	-51.77	0.72	-39	-50
Q14	-63.64	0.84	-66.13	0.92	-67	-61.17	0.78	-67.73	0.88	-69	-69.53
Q15	-88.12	1	-88.12	1	-66	-84.16	0.83	-84.84	0.93	-60	-88.12
Q16	-62.24	0.53	-62.3	0.68	-67	-49.41	0.63	-44.29	0.88	-67	-51.84
Q17	-63.11	0.7	-61.16	0.91	-67	-46.87	0.65	-51.69	0.88	-47	-56.18
Q18	-62.41	0.71	-75.68	0.69	-67	-74.28	0.55	-66.36	0.89	-67	-70.59
Q19	-63.13	0.74	-55.56	0.92	-66	-46.81	0.66	-50	0.89	-36	-57.14
Q20	-62.33	0.71	-64.71	0.69	-67	-55.55	0.61	-59.96	0.89	-78	-62.5

(b)

Rule base 1			Rule base 2		
Proposed method		Decision trees	Proposed method		Decision trees
Threshold 0.8	Threshold 0.9		Threshold 0.8	Threshold 0.9	
10.53 %	9.68 %	15.37 %	17.87 %	13.19 %	21.39 %

The two tested thresholds were chosen experimentally. The obtained results tend to show that there is an optimum threshold of 0.9. The choice of the threshold impacts the number and size of the groups obtained in the reconciliation step of the method. These must neither be too numerous and small (as in the case of a high threshold) nor too few and large (as in the case of a low threshold).

With rule base 1, the number of obtained groups was: 14 for threshold 0.8, and 45 for threshold 0.9. With rule base 2, the number of obtained groups was: 5 for threshold 0.8, and 25 for threshold 0.9. When the obtained groups are few and large (threshold 0.8), we can notice that the obtained predictions are more homogeneous among the tested queries. On the contrary, when the obtained groups are numerous and small (threshold 0.9), the obtained predictions are more various among the tested queries.

Figures 3 (a) and (b) present the error rates obtained query by query, with the proposed method for the optimum threshold 0.9, and with the decision tree method, respectively for rule base 1 and rule base 2.

We can make the following observation. The queries that obtained the most different results, if we compare both methods, are those for which exact or close answers were present in the rule base (such as for query *Q*9 in rule base 1), or those for which the closest answers, even if not so close, are quite different from the rest of the base and show different trends (such as for query *Q*10 in rule base 2).

The latter result is not very surprising since sensitivity to outliers is a well-known drawback of decision trees, and a strong point of our method which relies on the identification

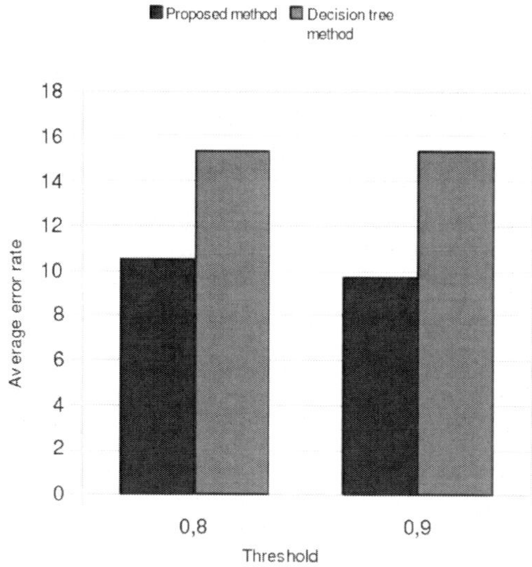

(a)

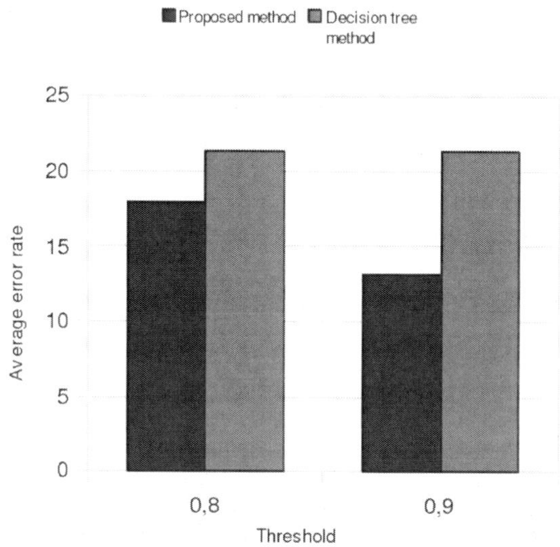

(b)

Figure 2. Average error rate for (a) rule base 1 and (b) rule base 2.

of common tendencies. Let us recall that our interest in (i) the case-based and (ii) the decision tree approaches is motivated, as previously mentioned, by specific features that are not handled by other methods (or not simultaneously), namely (i) missing values, (ii) both numerical and symbolic values and (iii) a limited number of cases (here rules). The proposed method thus takes the best from case-based and reconciliation approaches, moreover it is aware of ontological knowledge, and an improvement may legitimately be expected. Here we can note that the decision tree strategy, which processes step by step by considering each variable separately, turns out to be less relevant than the proposed method, which considers the rules globally, involving all the variables.

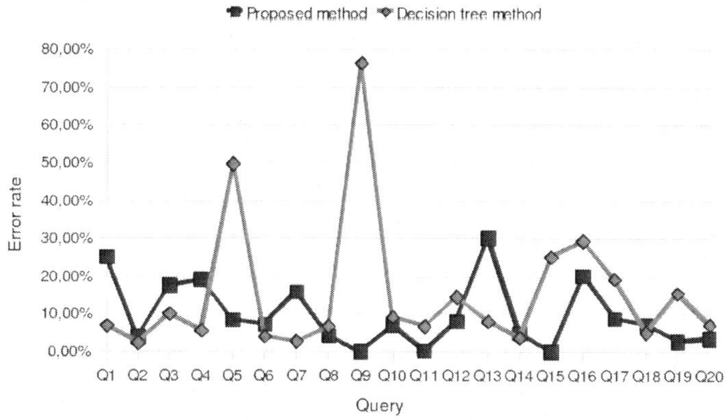

(a)

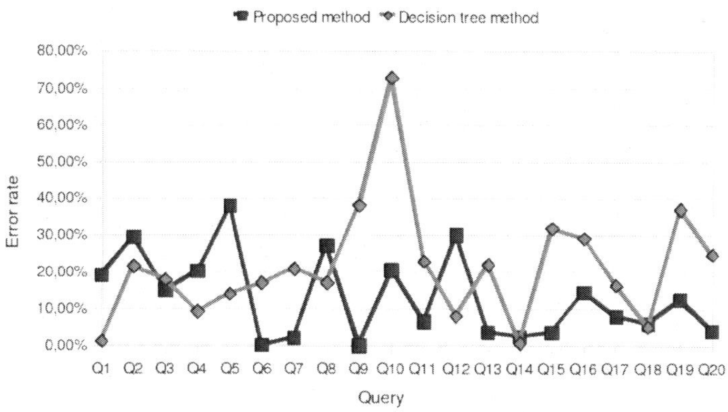

(b)

Figure 3. Error rates obtained for each query with (a) rule base 1 and (b) rule base 2.

7. Conclusion

This chapter focused on an approach to generate prediction rules relying on case-based and reconciliation methods, using an ontology.

The approach is particularly suitable to experimental sciences, to make predictions concerning unexplored experimental conditions. Since only a limited part of the huge number of possible experimental conditions have been explored and established as domain rules, knowledge bases are incomplete by nature. A solution consists in exploiting existing rules that concern close – although not identical – experimental conditions, as allowed by the presented approach.

Not every predictive method may be used in the considered context. Experimental conditions have missing values (not all the parameters are described in each rule), use both quantitative and qualitative parameters (numerical and symbolic values), and are scarce since we are not in a high speed data application but in a scarce-knowledge context with only a few hundreds of rules available. Few methods are able to deal with all these three issues and they basically are case-based approaches or decision trees methods. This explains why we decided to compare our work against these approaches.

Rule reconciliation showed to have advantages from a semantic and from a computational point of view. From a semantic viewpoint, it constitutes a consolidation of the knowledge represented by the set of rules, since it highlights the expression of common experimental tendencies. From a computational viewpoint, it speeds up the performances of the method by providing a restriction of the search space. The use of ontological knowledge both participates in performance improvement, through the disjunction relation in particular, and allows taking into account variable relevance, through identified functional dependencies. The experimentation on the application domain of cereal food design has shown a real potential of the approach. Compared with the results obtained by decision tree prediction, the proposed approach is more complete and more accurate.

To deal with information incompleteness which appears in the condition part of the rules, different ways may be followed during the similarity computation between two rules: (i) by considering, in the common description, only the filled properties in both rules. This means that the missing values in one of the two rules are ignored and do not participate in the similarity computation; (ii) by considering, in the common description, the whole description, i.e. the filled and the unfilled properties. In this case, the missing values are exploited as a negative information since they decrease the similarity scores. It will be worth studying more deeply the impact of these choices on the prediction results.

A complementary ongoing work deals with the design and validation of a domain expertise. It aims at the cooperation of two kinds of information, heterogeneous by their granularity levels and their formalisms: expert statements expressed in a knowledge representation model and experimental data represented in the relational model. In such a framework, the predictive approach presented in this chapter may be usefully introduced, in order to validate or invalidate the obtained predictions.

References

[1] Agnar Aamodt. Knowledge-intensive case-based reasoning and sustained learning. In *ECAI*, pages 1–6, 1990.

[2] Agnar Aamodt. Knowledge-intensive case-based reasoning in creek. In Peter Funk and Pedro A. González-Calero, editors, *ECCBR*, volume 3155 of *LNCS*, pages 1–15. Springer, 2004.

[3] Agnar Aamodt and Enric Plaza. Case-based reasoning: Foundational issues, methodological variations, and system approaches. *AI Commun.*, 7(1):39–59, 1994.

[4] Manuel Atencia, Jérôme David, and François Scharffe. Keys and pseudo-keys detection for web datasets cleansing and interlinking. In *EKAW*, pages 144–153, 2012.

[5] J. Baranyi and M.L. Tamplin. A new international effort for the advance of predictive microbiology, ComBase: Combined database of microbial responses to food environments. In *ICPMF'03*, pages 50–51, Quimper, France, June 2003.

[6] Mustafa Bilgic, Louis Licamele, Lise Getoor, and Ben Shneiderman. D-dupe: An interactive tool for entity resolution in social networks. In Patrick Healy and Nikola S.

Nikolov, editors, *International Symposium on Graph Drawing*, volume 3843 of *Lecture Notes in Computer Science*, pages 505–507. Springer, September 2005.

[7] Jean-Rémi Bourguet, Rallou Thomopoulos, Marie-Laure Mugnier, and Joël Abécassis. An artificial intelligence-based approach to deal with argumentation applied to food quality in a public health policy. *Expert Systems With Applications*, 40(11):4539–4546, 2013.

[8] Leo Breiman. *Classification and Regression Trees.* Boca Raton: Chapman & Hall/CRC, 1984.

[9] S. Choi, S. Cha, and C.C. Tappert. A survey of binary similarity and distance measures. *Journal of Systemics, Cybernetics and Informatics*, 8:43–48, 2010.

[10] William W. Cohen, Pradeep Ravikumar, and Stephen E. Fienberg. A comparison of string distance metrics for name-matching tasks. In *Proceedings of IJCAI-03 Workshop on Information Integration*, pages 73–78, August 2003.

[11] Sylvie Coste-Marquis, Caroline Devred, Sébastien Konieczny, Marie-Christine Lagasquie-Schiex, and Pierre Marquis. On the merging of dung's argumentation systems. *Artif. Intell.*, 171(10-15):730–753, 2007.

[12] Madalina Croitoru, Rallou Thomopoulos, and Nouredine Tamani. A practical application of argumentation in french agrifood chains. In *Information Processing and Management of Uncertainty in Knowledge-Based Systems - 15th International Conference, IPMU 2014, Montpellier, France, July 15-19, 2014, Proceedings, Part I*, pages 56–66, 2014.

[13] G. Dalbon, D. Grivon, and M.A. Pagnani. Continuous manufacturing process. In J.E. Kruger, R.B. Matsuo, and J.W. Dick, editors, *Pasta and noodle technology*. AACC, St Paul (MN-USA), 1996.

[14] Mathieu d'Aquin, Jean Lieber, and Amedeo Napoli. Case-based reasoning within semantic web technologies. In Jérôme Euzenat and John Domingue, editors, *AIMSA*, volume 4183 of *LNCS*, pages 190–200. Springer, 2006.

[15] Xin Dong, Alon Halevy, and Jayant Madhavan. Reference reconciliation in complex information spaces. In *ACM SIGMOD*, pages 85–96. ACM Press, 2005.

[16] Didier Dubois and Henri Prade. On the use of aggregation operations in information fusion processes. *Fuzzy Sets and Systems*, 142(1):143–161, 2004.

[17] M Ehrig. Similarity for ontologies - a comprehensive framework. In *in 13th European Conference on Information Systems*, 2005.

[18] Ahmed K. Elmagarmid, Panagiotis G. Ipeirotis, and Vassilios S. Verykios. Duplicate record detection: A survey. *IEEE Transactions on Knowledge and Data Engineering*, 19:1–16, 2007.

[48] Fatiha Saïs and Rallou Thomopoulos. Ontology-aware prediction from rules: A reconciliation-based approach. *Knowledge-Based Systems*, 67:117–130, 2014.

[49] Francois Scharffe, Yanbin Liu, and Chuguang Zhou. Rdf-ai: an architecture for rdf datasets matching, fusion and interlink. In *Proc. IJCAI 2009 workshop on Identity, reference, and knowledge representation (IR-KR), Pasadena (CA US)*, 2009.

[50] Christoph Schmitz, Andreas Hotho, Robert Jäschke, and Gerd Stumme. Content aggregation on knowledge bases using graph clustering. In York Sure and John Domingue, editors, *ESWC*, volume 4011 of *LNCS*, pages 530–544. Springer, 2006.

[51] G. Stumme, A. Hotho, and B. Berendt. Semantic web mining: State of the art and future directions. *J. of Web Semantics*, 4:124–143, 2006.

[52] Danai Symeonidou, Vincent Armant, Nathalie Pernelle, and Fatiha Saïs. SAKey: Scalable Almost Key discovery in RDF data. In *Proceedings of the 13th International Semantic Web Conference (ISWC2014)*, ISWC '14, page 16. Springer Verlag, 2014.

[53] Rallou Thomopoulos, Jean-François Baget, and Ollivier Haemmerlé. Conceptual graphs as cooperative formalism to build and validate a domain expertise. In *ICCS*, pages 112–125, 2007.

[54] Rallou Thomopoulos, Madalina Croitoru, and Nouredine Tamani. Decision support for agri-food chains: A reverse engineering argumentation-based approach. *Ecological Informatics*, 2014.

[55] Rallou Thomopoulos, Sébastien Destercke, Brigitte Charnomordic, Iyan Johnson, and Joël Abécassis. An iterative approach to build relevant ontology-aware data-driven models. *Information Sciences*, 221:452–472, 2013.

[56] Julius Volz, Christian Bizer, Martin Gaedke, and Georgi Kobilarov. Discovering and maintaining links on the web of data. In *Proceedings of the 8th International Semantic Web Conference(ISWC)*, ISWC '09, pages 650–665, Berlin, Heidelberg, 2009. Springer-Verlag.

[57] Daisy Zhe Wang, Xin Luna Dong, Anish Das Sarma, Michael J. Franklin, and Alon Y. Halevy. Functional dependency generation and applications in pay-as-you-go data integration systems. In *WebDB*, 2009.

[58] William E Winkler. Overview of record linkage and current research directions. Technical report, Statistical Research Division U.S. Census Bureau Washington, DC 20233, 2006.

[59] Zhibiao Wu and Martha Palmer. Verbs semantics and lexical selection. In *Proceedings of the 32nd annual meeting on Association for Computational Linguistics*, pages 133–138, Morristown, NJ, USA, 1994. Association for Computational Linguistics.

[60] L.S. Young. Application of Baking Knowledge in Software Systems. In *Technology of Breadmaking - 2nd edition*, pages 207–222. Springer, US, 2007.

[61] J. Zhang, A. Silvescu, and V. Honavar. Ontology-driven induction of decision trees at multiple levels of abstraction. *Lecture Notes in Computer Science*, pages 316–323, 2002.

In: Case-Based Reasoning
Editor: Tao Lin
ISBN: 978-1-63482-952-6

Chapter 2

Application of Case-Based Reasoning and Group Decision-Making Methods for Planning in a Case of Failures of Technical Systems

Aleksandr F. Berman, Olga A. Nikolaichuk, Galina S. Maltugueva and Aleksandr Yu. Yurin*
Matrosov Institute for Systems Dynamics and Control Theory, Siberian Branch Russian Academy of Sciences, Irkutsk, Russia

Abstract

The chapter describes the joint application of case-based reasoning and group decision-making methods for planning of measures in a case of technical system failures. The detailed description of the proposed joint application is presented. The described approach for planning can be considered as a method of a transformational adaptation in the context of planning. In planning, cases contain the solutions (as plans) in the form of completely or partially ordered sequences of actions (measures). Transformation operators transform existing plans (from the case base) into new ones that are reused in new problem situations. In the proposed approach, the order of solution parts is determined with the aid of strictly formalized decision-making methods, using proven mathematical methods of a decision theory, and corresponds to the 'group preferences' concept of a decision theory, i.e., it has the transitivity property. The chapter also describes the case model; case retrieval and case reuse algorithms; some components of software and a case study.

Keywords: case-based reasoning, group decision-making, expert system, repeated failure, recommendations, prevention, planning

* Corresponding author: Irkutsk, Russia, Lermontov st., 134, Email: iskander@icc.ru.

INTRODUCTION

One of the important problems in many subject domains (including automated planning) is the problem of improving the efficiency of experience reuse in solving new problems. Case-based reasoning [1-3] is an approach to solving this problem. This approach uses (or simulates) a human reasoning model in decision-making by analogy and does not require a deep analysis of a subject domain. An effective application of this approach can be made on the basis of superficial knowledge represented in the form of structured patterns or cases.

Case-based decision-making as a separate scientific area in the field of knowledge-based systems is oriented primarily toward decision-making through the reuse of experience. This orientation requires investigation and development of effective new methods for a representation (modelling), evaluation, storage, indexing, retrieval and adaptation of actual knowledge. In accordance with this methodology, the decision-making process classically consists of four stages [1-3]:

- Retrieval of similar cases, including recognition of a new problem presented in the form of a pattern (a case);
- Reuse of information from retrieved cases, including its adaptation (if necessary);
- Revision of a new solution for correctness; and
- Retaining a new case in a case-base.

The reuse stage is the most complex because it utilises deep subject-domain knowledge. This knowledge allows us to estimate qualitative differences between retrieved cases and the new problem and to take into account these differences in the formation of the final solution, i.e., to adapt the solutions received.

The problem of developing adaptation methods has remained constant since the early investigations into case-based reasoning [4-7]. Case adaptation methods can be classified [1, 2, 8-10] into two basic categories: transformational analogy and derivational analogy.

In transformational analogy, case solutions are reused in new situations by making appropriate changes, if necessary with the aid of transformation operators. Transformational analogy does not consider how the reused solution was originally obtained; it is only concerned with the solution itself. In derivational analogy [11], solutions contain derivational traces that can be replayed relative to the new problem (and do not require transformational operators). However, derivational analogy requires the derivational trace to be known. When this is not possible, transformational analogy is the better choice because the solution itself can be used for adaptation.

This chapter describes an approach that can be considered a method of transformational adaptation in the context of planning [8, 10, 12]. In planning, cases contain solution plans for previous problems in the form of a completely or partially ordered sequence of actions. These plans are reused in the new situation by making suitable changes where appropriate. Transformation operators transform existing plans into new ones.

It is proposed to implement the transformation operator on the basis of group decision-making (GDM) methods [13, 14]. The final solution is the result of aggregating multiple solutions of similar cases rather than a single best-case solution. Parts of solutions coming from different retrieved cases are combined together in the process of aggregation. An ordered sequence of parts corresponds to the concept 'group preference' [13] of a decision theory.

Problem Statement

The feature of case-based reasoning [1, 2] is to obtain a set of similar cases from the case retrieval stage. The cases are ordered according to their similarity, and a decision-maker creates a solution for the new situation on the basis of one or more similar cases. In the context of planning, the solution is a plan in the form of a sequence of actions ordered by priority or acceptance time. Thus, a decision-making process (in particular, a procedure combining parts of solutions of similar cases) can be improved with the aid of GDM methods [13-15].

The use of GDM methods provides for obtaining the solution by aggregating solutions from similar cases, even those that are usually excluded from a decision-making process.

Consider a formal statement of the problem.

It is given: $C = \{c_1,\ldots,c_M\}$ is a set of retrieved cases, and each case is described by the triple: $c_i = \langle p_i, d_i, s_i \rangle$, where p_i is an *i*-case description, d_i is an *i*-case solution, and s_i is an *i*-case similarity, $i \in \overline{1,M}$.

Consider the situation where case solution d_i is a set (or a sequence) of alternatives (events, actions) ordered by decision time, then $d_i = \langle A_i^c, d_i^Q \rangle$ where $A_i^c = a_{i1}^{h_1},\ldots,a_{iZ_i}^{h_{Z_i}}$ is the set of alternatives (events, actions), and $A_i^c \cap A_j^c \neq \varnothing$. The union of sets A_i^c forms a set of all possible alternatives A for C: $A = \bigcup_{i=1}^{M} A_i^c = a^1,\ldots,a^H$, $H = |A|$, and $h_z \in \overline{1,H}$. $d_i^Q = a_{i1}^{h_1} Q_1^i \ldots Q_{Z_i-1}^i a_{iZ_i}^{h_{Z_i}}$ is the sequence of alternatives (or ranking), where $Q \in \{\succ, \approx\}$, i.e., Q is a relation of strict preference ($\succ$) or equivalence ($\approx$) [13], $Z_i = |A_i^c|$, and c is an index of affiliation of alternatives to cases. Thus, it is possible that $A_i^c = A_j^c$, but $d_i^Q \neq d_j^Q$, $i,j \in \overline{1,M}$.

We must obtain the solution d^* by aggregating the solutions of cases from C and taking into account their similarities.

The solution d^* corresponds to the concept 'group preference'. The individual preference is a formal description of an expert's ability to compare (or to order) different alternatives in the set A and can be described quantitatively (by units of measurement) or qualitatively (using the expert's subjective opinion). In this work the preference is described qualitatively, and can include of strict preference ($\succ$) and equivalence ($\approx$). The equivalence relation ($\approx$) is a binary relation on the set A that satisfies the following conditions:

- reflexivity: $\forall a^i \in A$, $i \in H$: $a^i \approx a^i$;
- symmetry: if $a^i \approx a^j$ then $a^j \approx a^i$, $\forall a^i, a^j \in A$, $i,j \in H$; and
- transitivity: if $a^i \approx a^j$ and $a^j \approx a^k$ then $a^i \approx a^k$, $\forall a^i, a^j, a^k \in A$, $i,j,k \in H$.

The strict preference relation ($\succ$) is a binary relation on the set A that satisfies the following conditions:

- irreflexivity: $\neg\exists a^i \in A$, $i \in H$; $a^i \succ a^i$;
- transitivity: $a^i \succ a^j \wedge a^j \succ a^k \Rightarrow a^i \succ a^k$ $a^i \approx a^j \wedge a^j \succ a^k \Rightarrow a^i \succ a^k$; and $a^i \succ a^j \wedge a^j \approx a^k \Rightarrow a^i \succ a^k$, $a^i \approx a^k$, $\forall a^i, a^j, a^k \in A$, $i, j, k \in H$.

Thus, the group preference is a formal consensus of experts working in a group, and each of them has an individual preference.

The Joint Application of Case-Based Reasoning and Group Decision-Making Methods

The joint application of case-based reasoning and group decision-making methods allows to represent information of the solutions of retrieved cases in the form of datasets. The obtained datasets are processed by several GDM methods with the following coordination (or aggregation) of results.

The generalised algorithm of the proposed approach consists of the following steps:

Step 1: Case selection (retrieval) for forming datasets;
Step 2: Dataset formation for processing by GDM methods;
Step 3: Dataset processing by GDM methods; and
Step 4: Obtaining a final aggregated solution by the aggregation of intermediate solutions obtained in Step 3.

Let us consider the steps of the algorithm in detail.

Step 1: Case selection is based on the results of the case-retrieval procedure. There are many methods and modifications that can be used for case retrieval [1-3]: nearest neighbour, decision trees, etc. The most popular method is the nearest neighbour, based on an assessment of similarity with the aid of different metrics, for example, Euclidean, City-Block-Metric, etc. The proposed approach uses the nearest neighbour method and Zhuravlev metric [16] with normalisation:

$$d_G(\bar{x}, \bar{y}) = \sum_{i=1}^{N} w_i h_G(x_i, y_i) \Big/ N, \quad h_G(x_i, y_i) = \begin{cases} \text{for quantitative} \begin{cases} 1, if |x_i - y_i| < \xi \\ 0, else \end{cases} \\ \text{for qualitative} \begin{cases} 1, x_i = y_i \\ 0, x_i \neq y_i \end{cases} \end{cases},$$

where w_i is the information weight, and ξ is the constraint on the difference between the values of properties.

At the same time, the normalisation (or standardization) procedure is performed to bring all values onto a single scale:

$$x_{ik} \rightarrow \left(x_{ik} - \min_k x_{ik} \right) \Big/ \left(\max_k x_{ik} - \min_k x_{ik} \right).$$

In most cases, the case selection (to obtain a final solution for the new problem) is a subjective process. That is why the implementation of this approach in [17] allows the user to determine the confidence interval for similarity to select the cases. In this illustrative example (see below), cases with similarity greater than or equal to 0.5 were selected (the confidence interval is [0.5;1]).

The result of Step 1 is the set of cases $C^* = c_1,\ldots,c_m$, where $c_i = \langle p_i, d_i, s_i \rangle$, $i = \overline{1,m}$, $m \le M$, $s_i \in [0.5;1]$, $d_i = \langle A_i^c, d_i^Q \rangle$, $A_i^c = a_i^{h_1},\ldots,a_{iZ_i}^{h_{Z_i}}$, and $d_i^Q = a_{i1}^{h_1} Q_1^i \ldots Q_{Z_i-1}^i a_{iZ_i}^{h_{Z_i}}$.

Step 2. Dataset formation: T on the basis of information from cases that are selected in Step 1. In this step, the following tasks are solved:

2.1. Grouping cases by taking into account alternatives of their solutions; and
2.2. Taking into account the similarity of cases.

The first task is solved as follows: the set of cases C^* selected in Step 1 is divided into disjoint subsets $\overline{C}_n$, $C^* = \bigcup_{n=1}^{q} \overline{C}_n, q \le m$, $\overline{C}_{n1} \cap \overline{C}_{n2} = \varnothing$, and $n1 \ne n2, \forall n1, n2 \in \overline{1,q}$, each of which represents a set of cases $\overline{C}_n = c_{n1},\ldots,c_{nK_n}$, $n = \overline{1,q}$. In addition, each subset $\overline{C}_n$ has a unique set of alternatives, i.e., sets of alternatives are equal for cases from one subset: $\forall c_{nk1}, c_{nk2} \in \overline{C}_n : A_{nk1}^c = A_{nk2}^c, k1 \ne k2$; and $k1, k2 \in \overline{1, K_n}$, and sets of alternatives are different for cases from different subsets: $\forall c_{n1k1} \in \overline{C}_{n1}, \forall c_{n2k2} \in \overline{C}_{n2} : A_{n1k1}^c \ne A_{nk2}^c$; $n1 \ne n2$; $n1, n2 \in \overline{1,q}$; $k1 \ne k2$; and $k1, k2 \in \overline{1, K_n}$. The dataset is formed for each subset $\overline{C}_n \rightarrow T_n$, $n = \overline{1,q}$, so the result of Step 2 is the set of datasets $T = T_1,\ldots,T_q$.

Let us describe components of the dataset $T_n = \langle A_n^T, R_n, S_n \rangle$, $n = \overline{1,q}$, $q \ge 1$, where A_n^T is a set of alternatives to the dataset: $A_n^T = a_{n1}^{h_1},\ldots,a_{nL_n}^{h_{L_n}}$, L_n is a number of alternatives in the dataset, and R_n is a set of rankings (or sequences of alternatives), $R_n = \langle d_{n1}^Q, e_{n1} \rangle, \langle d_{n2}^Q, e_{n2} \rangle, \ldots, \langle d_{nr}^Q, e_{nr} \rangle$, where r is a number of unique solutions (rankings), $r \le m$, and d_{nr}^Q is a case solution (represented in the form of a ranking) that is identical for a number of cases e_{nr}. S_n is an informational weight of the dataset T_n. Calculation of the informational weight allows us to solve the second task of Step 2. It is equal to the mean expression of similarities for cases of the dataset.

Thus, the components of a case are mapped into the components of a dataset: $\{A_{nk}^c\}_k \to A_n^T$, $\{d_{nk}^Q\} \to R_n$, $\{s_{nk}\} \to S_n$.

Let us consider an example illustrating the first two steps of the generalised algorithm for the proposed approach.

The set of cases C is given, and the cases are ordered according to their similarities:

$$c_1 = \langle p_1, d_1, s_1 \rangle,\ A_1^c = \ a_{11}^1, a_{12}^2, a_{13}^3 \ ,\ d_1^Q = a_{11}^1 \succ a_{12}^2 \succ a_{13}^3,\ s_1 = 1.0$$

$$c_2 = \langle p_2, d_2, s_2 \rangle,\ A_2^c = \ a_{21}^1, a_{22}^2, a_{23}^3 \ ,\ d_2^Q = a_{21}^1 \succ a_{22}^2 \succ a_{23}^3, s_2 = 1.0$$

$$c_3 = \langle p_3, d_3, s_3 \rangle,\ A_3^c = \ a_{31}^1, a_{32}^3, a_{33}^2 \ ,\ d_3^Q = a_{31}^1 \succ a_{32}^3 \succ a_{33}^2, s_3 = 0.8$$

$$c_4 = \langle p_4, d_4, s_4 \rangle,\ A_4^c = \ a_{41}^1, a_{42}^3, a_{43}^4 \ ,\ d_4^Q = a_{41}^1 \succ a_{42}^3 \succ a_{43}^4, s_4 = 0.7$$

$$c_5 = \langle p_5, d_5, s_5 \rangle,\ A_5^c = \ a_{51}^1, a_{52}^3, a_{53}^4 \ ,\ d_5^Q = a_{51}^1 \succ a_{52}^3 \succ a_{53}^4,\ s_5 = 0.6$$

$$c_6 = \langle p_6, d_6, s_6 \rangle,\ A_6^c = \ a_{61}^3, a_{62}^4 \ ,\ d_6^Q = a_{61}^3 \succ a_{62}^4, s_6 = 0.6$$

$$c_7 = \langle p_7, d_7, s_7 \rangle,\ A_7^c = \ a_{71}^3, a_{72}^2, a_{73}^1 \ ,\ d_7^Q = a_{71}^3 \succ a_{72}^2 \succ a_{73}^1, s_7 = 0.5$$

$$c_8 = \langle p_8, d_8, s_8 \rangle,\ A_8^c = \ a_{81}^2, a_{82}^3, a_{83}^1 \ ,\ d_8^Q = a_{81}^2 \succ a_{82}^3 \succ a_{83}^1, s_8 = 0.5$$

$$c_9 = \langle p_9, d_9, s_9 \rangle, \ldots, s_9 = 0.3$$

$$c_{10} = \langle p_{10}, d_{10}, s_{10} \rangle, \ldots, s_{10} = 0.1$$

$$c_{11} = \langle p_{11}, d_{11}, s_{11} \rangle, \ldots, s_{11} = 0.0$$

...

$A = \ a^1, a^2, a^3, a^4$ is a set of possible alternatives.

The set $C^* = \{c_1, c_2, c_3, c_4, c_5, c_6, c_7, c_8\}$ is obtained from the selection of cases with similarities greater than or equal to 0.5 (Step 1). Three subsets of cases are formed as a result of the analysis of C^* (Step 2, Task 2.1): $C^* = \ \overline{C}_1, \overline{C}_2, \overline{C}_3$, $\overline{C}_1 = \ c_1, c_2, c_3, c_7, c_8$, $\overline{C}_2 = \ c_4, c_5$, and $\overline{C}_3 = \ c_6$. The set of datasets $T = \ T_1, T_2, T_3$ is obtained on the basis of these subsets (Step 2):

$$T_1 = \langle A_1^T, R_1, S_1 \rangle:\ A_1^T = \ a_{11}^1, a_{12}^2, a_{13}^3 \ ,\ R_1 = \ \langle d_{11}^Q, 2 \rangle, \langle d_{12}^Q, 1 \rangle, \langle d_{13}^Q, 1 \rangle, \langle d_{14}^Q, 1 \rangle \ ,$$

$$d_{11}^Q = a_{111}^1 \succ a_{112}^2 \succ a_{113}^3,\ d_{12}^Q = a_{121}^1 \succ a_{122}^3 \succ a_{123}^2,\ d_{13}^Q = a_{131}^3 \succ a_{132}^2 \succ a_{133}^1,\ d_{14}^Q = a_{141}^2 \succ a_{142}^3 \succ a_{143}^1,$$

$S_1 = 0.76$;

$$T_2 = \langle A_2^T, R_2, S_2 \rangle : A_2^T = \ a_{21}^1, a_{22}^3, a_{23}^4 \ , R_2 = \ \langle d_{21}^Q, 2 \rangle \ ,\ d_{21}^Q = a_{211}^1 \succ a_{212}^3 \succ a_{213}^4,\ S_2 = 0.65;$$

and

$$T_3 = \langle A_3^T, R_3, S_3 \rangle:\ A_3^T = \ a_{31}^3, a_{32}^4 \ ,\ R_3 = \ \langle d_{31}^Q, 1 \rangle \ ,\ d_{31}^Q = a_{311}^3 \succ a_{312}^4,\ S_3 = 0.6.$$

Step 3: Processing datasets from T by GDM methods

Thus, the set of datasets $T = \; T_1,\ldots,T_q$ is obtained in Step 2.

We must obtain the set of solutions $D = \; d_1^*,\ldots,d_q^*$ for T where $d_n^* = a_{n1}^{h_1} Q_1^n \ldots Q_{l-1}^n a_{nL_n}^{h_{l_n}}$ is a group preference relation represented in the form of a ranking.

The efficiency of dataset processing can be improved as follows:

- by processing each dataset with the aid of several GDM methods in coordination with the results. We assume that the application of several GDM methods reduces the probability of incorrect solutions and paradoxes;
- by applying an algorithm of the results' sequential coordination. This algorithm allows us to gradually obtain a solution in two stages. A certain set of GDM methods operates at the first stage, and its results are group preference (method selection also depends on the user's subjective preferences). Further, the results of the first stage are used as the input for the second stage, which uses only one GDM method. This algorithm provides a solution that does not depend on the features (or restrictions) of the various GDM methods.

The result of Step 3 is the set of intermediate solutions $D = \; d_1^*,\ldots,d_q^*$.

Step 4. Obtaining a final aggregated solution. The purpose of this step is to obtain the aggregated solution d^* by coordinating (or aggregating) the intermediate solutions from D. This step can be omitted if $|D| = 1$; otherwise, we must form the dataset T^* on the basis of D and process it with the aid of GDM methods. Thus, the intermediate solutions $d_n^*, n \in \overline{1,q}$ should be complemented by alternatives from A to form generalised rankings in R^* of identical capacity. All complemented alternatives are equivalent to each other.

The result of Step 4 is the final aggregate solution d^*.

Let us explain Step 4 in this example. The following intermediate solutions for $T = \; T_1,T_2,T_3$ are obtained in Step 3: $d_1^* = a_{11}^1 \succ a_{12}^2 \succ a_{13}^3$, $d_2^* = a_{21}^1 \succ a_{22}^3 \succ a_{23}^4$, and $d_3^* = a_{31}^3 \succ a_{32}^4$.

Because $|D| = 3$ the intermediate solutions are complemented by alternatives from $A = \bigcup_{n=1}^{3} A_n^T = \; a^1,a^2,a^3,a^4$, and as a result a new set of rankings is obtained: $R^* = \; \langle d_1^{Q*}, e_1^* \rangle, \langle d_2^{Q*}, e_2^* \rangle, \langle d_3^{Q*}, e_3^* \rangle$, where $d_1^{Q*} = a_{11}^1 \succ a_{12}^2 \succ a_{13}^3 \succ a_{14}^4$, $e_1^* = 76$; $d_2^{Q*} = a_{21}^1 \succ a_{22}^3 \succ a_{23}^4 \succ a_{24}^2$, $e_2^* = 65$; and $d_3^{Q*} = a_{31}^3 \succ a_{32}^4 \succ a_{33}^1 \approx a_{34}^2$, $e_3^* = 60$. e_n^* is formed by the transformation of an informational dataset weight: $S_q \rightarrow e_q^*$, $q = \overline{1,n}$, $e_q^* = S_q * 100$ $\forall q = \overline{1,n}$. This transformation allows us to move from fractional numbers to integers that are used to describe generalised rankings.

The final result is $d^* = a_1^1 \succ a_2^3 \succ a_3^2 \approx a_4^4$.

SOFTWARE FOR PLANNING IN A CASE OF FAILURES OF TECHNICAL SYSTEMS

The problem of the effective definition of recommendations for a prevention of repeated failures requires the development of intelligent software. Such software in the form of the expert system [18] has been developed for solve problems in the safe exploitation of technical systems.

The software implements the following main functions:

- description of the undesirable state (an incident, a failure or emergency);
- identification of the cause-and-effect complex of factors, which condition the undesirable state;
- substantiation and planning the works and necessary measures for the prevention of repeated failures;
- identification and interpretation of incidents (failures, emergencies), and it's registration in the case-base;
- generation of specialized reports.

Application of the software requires the description of the object domain (the specificity of the enterprise): mechanical systems and devices (this information is collected in 'Equipment' and 'Materials' databases); incidents, failures and emergencies (formalized and collected as cases in the case-base).

The model of the case represents some compact description of knowledge about a problem situation, its decision, and contains the following two principal components: the component containing the description of the problem and the component of decision-making related to this problem. Consider the components of this case-based model:

$$\text{Case} = \langle \text{Problem, Decision} \rangle . \tag{1}$$

The component of the problem's description (Problem) includes the object's properties, which reflect its states on different stages of manufacturing and exploitation. These properties have been obtained in the process of investigation of the cases. In our case, the description of the problem is based on the aggregated model of the scrutinized object, which can be found in [19, 20].

$$\text{Problem} = \left\langle \text{O}^{\text{Stat}}, \text{O}^{\text{Dyn}} \right\rangle ,$$

where O^{Stat} – the static information model of the object (which includes information representing the object on the design stage); O^{Dyn} – the dynamic information model of the object (which includes information representing the object on the stage of its exploitation).

The static model (on the stage of design):

$$\text{O}^{\text{Stat}} = \left\langle \text{EI}^{\text{D}}, \text{P}^{\text{D}} \right\rangle ,$$

where EI^D – presumed factors of external influences; P^D – presumed object's properties: $P^D = \langle N_U, TechChar, TechR, TechC, P_S, P_R, TechState^D \rangle$, N_U – a number of unique elements inside the object; TechChar – technical characteristics; TechR – technical requirements; TechC – technical conditions; P_S – properties of the system of safety, for example, the availability and types of protective devices; P_R – properties of the system of reliability, for example, the availability and the kinds of embedded diagnostic systems; the level of confirmation of periodicity of the technical state assessment; $TechSate^D$ – the object's initial technical state, $TechSate^D = \{H_i\}$, H_i – i-mounting heredity, D – the index of design stage.

The dynamic model (on the stage of exploitation):

$$O^{Dyn} = \langle T, IF^E, P^E \rangle,$$

wher T – discrete time moments of the object's life cycle; IF^E – influencing factors related to the object's exploitation; P^E – exploitation properties of the object (this set corresponds to the structure of properties related to design); $TechSate^E$ – exploitation states of the object, $TechSate^E = \{UP_l\}$, UP_l – l-undesirable processes for the object; $TechState^E = \{TechState^k, TechState^k \rightarrow TechState^{k+1}\}$, $\rightarrow$ – the cause-effect relation; k – is the index used for the stage of undesirable processes, $k = \overline{1,4}$. At the moment of beginning of functioning of object T_0 , its design and its exploitation states coincide. This fact may be expressed by the following identity $TechState^D \equiv TechState^E$.

$$UP_l = \{UP_{lk} \mid UP_{lk} = \langle M_{lk}, K_{lk}, S_{lk} \rangle,\ M_{lk} = \langle EI^E, P^E \rangle,\ UP_{lk} \rightarrow UP_{lk+1}$$

$$M_{lk} \rightarrow \{K_{lk}^{mt}\}_m, \{K_{lk}^{mt}\}_m \rightarrow \{S_{lk}^{pt}\}_p\}$$

where UP_{lk} is the k-stage of the l-undesirable process; M_{lk} is the mechanism of the k-stage of the l- undesirable process (the set of object's properties and the set of factors, which influence the object); K_{lk} is the kinetics of the k-stage of the l-undesirable process (the set of events, which describe hazardous processes and phenomena); S_{lk} are the signs of the k-stage of the l-undesirable process; K_{lk}^{mt} is the m-event of the k-stage of the l-undesirable process at the time moment t; S_{lk}^{pt} is the p-signs of the k-stage of the l-undesirable process at the time moment t; l is the index of the undesirable process; m is the index of kinetics of the undesirable process; p is the index of the indicator of kinetics of the undesirable process; $\rightarrow$ is the sign of cause-effect relations.

The component of the case, which describes the decision related to the problem, includes the following elements:

$$Decision = \langle Cause, Controling\_actions \rangle,$$

where Cause is a description of the total complex of interconnected causes, which may lead (the stage of design) or have already led (the stage of exploitation) to some hazardous states; Controlling_actions represent a description of the sequence of some controlling decisions, which have been made to the end of preventing (the stage of design) or localization, liquidation and minimization of negative effects the stage of exploitation.

Consider the object-oriented model of case on the basis of (1) and the object-oriented model of concepts 'incident' [18] and 'failure', where the concept of 'incident' implies damage of a system, violations of some parameters and other deviations from the norm which may lead to emergency.

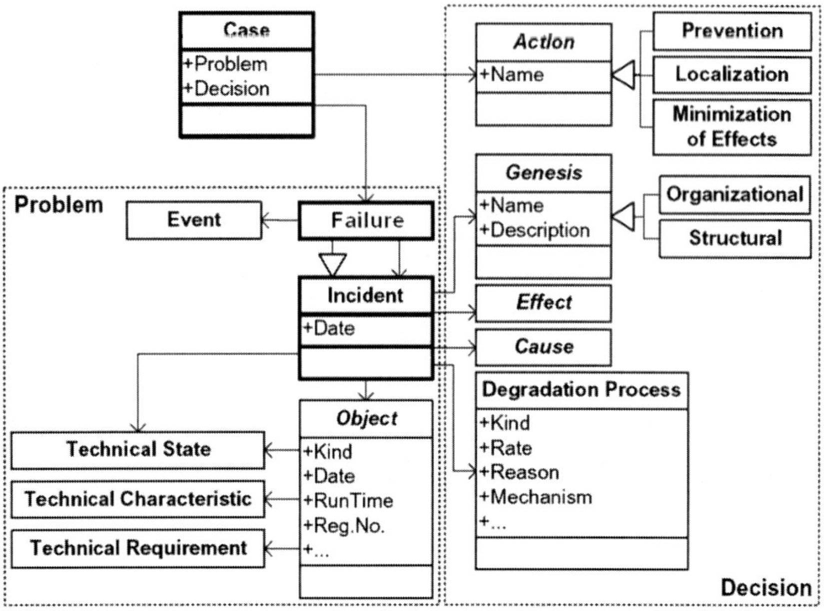

Figure 1. The case model [20].

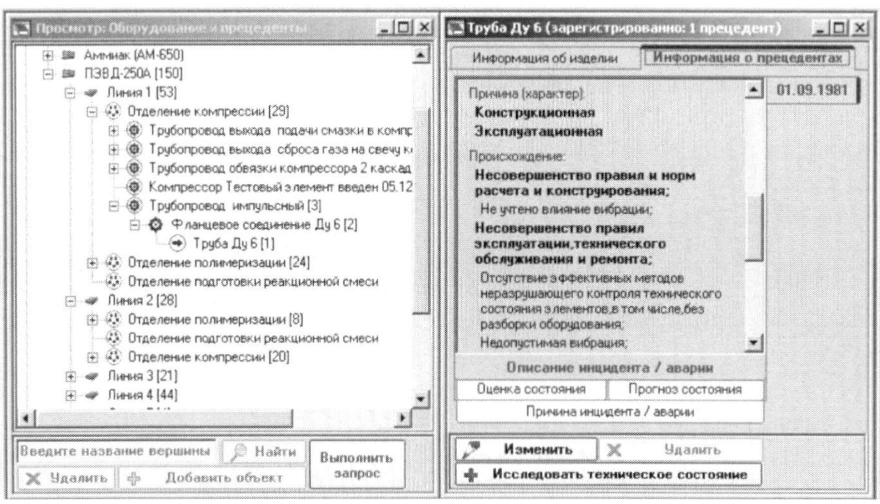

Figure 2. The GUI form of the software: The selection of the technical object.

The case model represented in Figure 1 describes a scenario of a failure. It presumes the following information: the name of the object and its hierarchical belonging (a technological complex – a technical system – a mechanical system); actual exploitation conditions; observed parameters related to variations of the technical state and their external manifestations; the sequence of states, which have led to the each hazardous stage, while including the properties of these states; causes of states; the structural genesis, i.e., the constructive element, which has become the cause of the hazardous state; the organization genesis, i.e., violations and imperfections, which have been admitted on all the stages of its life cycle; the decision made to control the undesirable state, effects (consequences) of undesirable processes, etc.

The software is used for decision support at planning in a case of failures of technical systems in the petrochemical industry [18] and for the investigation of safety and reliability in Matrosov Institute for System Dynamics and Control Theory, Siberian Branch of the Russian Academy of Sciences.

Let us describe an example of the software application.

The first step: description of the incident. The user can select the technical object (Figure 2) and to define the diagnostic features of the current technical state, including:

the signs - crack (cracks);

- parameters of the sign:
    - the source of damage – the stress concentrator,
    - the location - on the outer surface,
    - the specialty of the location – the threaded surface.

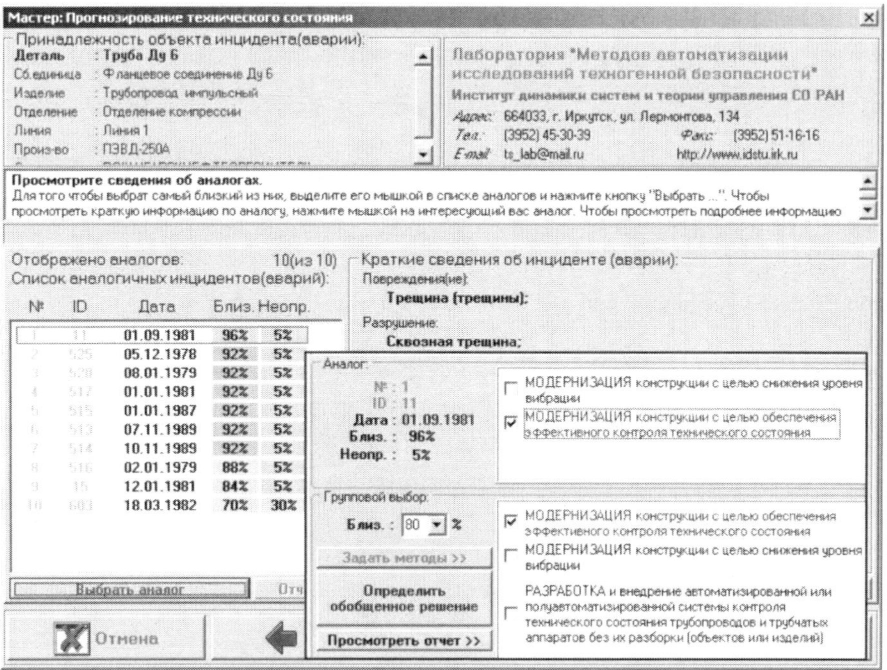

Figure 3. The GUI form of the software: Preview and revision of results of the case retrieval and GDM.

The second step: the case retrieval.

The third step: the preview and revision of results of the case retrieval (Figure 3). In particular, 10 cases are retrieved. The best case is №1 (ID 11) which describes the failure

(failure date is 01.09.1981) of the pipe with the angular diameter of 6 mm. This pipe is a part of the flange connection of the pulse conduit which are located in the compression section. The case №1 has the maximal similarity - 96% and the minimal uncertainty - 5%.

A decision-maker can review descriptions of cases retrieved and make a decision on the basis of his own experience and knowledge. The decision of the case №1 and results of the application of the proposed approach are represented on Figure 3. Decisions of cases with similarity in the range (80%; 100%) were chosen as initial data for GDM procedure.

CONCLUSION

This chapter describes an approach to improving the efficiency of case-based decision support with the aid of group decision-making methods. The proposed approach can be considered a method of a transformational adaptation in the context of planning. In planning, cases contain the solutions (as plans) for previous problems in the form of completely or partially ordered sequences of actions (parts of a solution). Transformation operators transform existing plans into new ones that are reused in new situations. In the proposed approach the transformation operator is implemented on the basis of GDM methods.

The advantages of this approach:

- the transformation operator takes into account multiple solutions of the best cases;
- the order of solution parts is determined with the aid of strictly formalized decision-making methods using proven mathematical methods of a decision theory; and
- the order of solution parts corresponds to the 'group preference' concept of a decision theory, i.e., it has the transitivity property.

The limitations of this approach:

- it is applicable only for planning, when solutions are ordered sequences of actions; and
- the final solution is just 'average', i.e., it is not always acceptable.

The practical significance of the proposed approach is determined by the GDM – the opportunity to establish a more accurately ordered sequence by taking into account all points of view (in our case, an ordered sequence is a solution). The best and worst elements in the sequence are usually obvious, but revealing the distribution of intermediates can be difficult.

The proposed approach has been applied for decision support in planning in the case of failures of technical systems. In particular, new module that implements this approach has been added to intelligent software [18], and the application of the proposed approach is described in detail in [17]. The case model includes a description of diagnostic signs of an incident or failure. These signs allow us to identify the current technical state of the object of the investigation. In addition, the case model includes a list of actions (or recommendations) to prevent and reduce losses from failures. Actions (or recommendations) form a strictly ordered sequence or a plan. The proposed approach expands the recommendation set to prevent repeated failures. This expansion is achieved by integrating solution elements of retrieved cases that were previously ignored.

REFERENCES

[1] Aamodt, A. & Plaza, E. (1994). Case-based reasoning: foundational issues, methodological variations, and system approaches. *AI Communications.*, *7*(1), 39-59.

[2] Bergmann, R. (2002). Experience Management: Foundations, Development Methodology and Internet-Based Applications. *Lecture Notes on Artificial Intelligence, 2432*, 1-393.

[3] De Mantaras, L. R., Mcsherry, D., Bridge, D., Leake, D., Smyth, B., Craw, S., Faltings, B., Maher, M. L., Cox, M. T., Forbus, K., Keane, M., Aamodt, A. & Watson, I. (2005). Retrieval, reuse, revision and retention in case-based reasoning. *Knowledge Engineering Review*, *20*(3), 215–240.

[4] Kolodner, J. L., Morgan Kaufmann, & San Mateo. (1993). Case-Based Reasoning, CA.

[5] McSherry, D. (1998). An adaptation heuristic for case-based estimation. *LNCS*, *1488*, 184-195.

[6] Plaza, E. & Arcos, J. L. (2002). Constructive adaptation. *Lecture Notes on Artificial Intelligence*, *2416*, 306-320.

[7] Wilke, W., Smyth, B. & Cunningham, P. (1998). Using configuration techniques for adaptation. *LNCS*, *1400*, 139-168.

[8] Cox, M. T., Muñoz-Avila, H. & Bergmann, R. (2006). Case-based planning. *Knowledge Engineering Review*, *20*(3), 283-287.

[9] Mitra, R. & Basak, J. (2005). Methods of case adaptation: A survey. *International Journal of Intelligent Systems*, *20* (6), 627-645.

[10] Muñoz-Avila, H. & Cox, M. T. (2008). Case-Based Plan Adaptation: An Analysis and Review. *IEEE Intelligent Systems*, *23*(4), 75-81.

[11] Veloso, M. M. & Carbonell, J. G. (1993). Derivational Analogy in PRODIGY: Automating Case Acquisition, Storage, and Utilization. *Machine Learning*, *10*(3), 249-278.

[12] Britanik, J. & Marefat, M. (2004). CBPOP: A Domain-Independent Multi-Case Reuse Planner. *Computational Intelligence*, *20*(2), 405-443.

[13] Arrow, K. J., Sen, A. K. & Suzumura, K. (2002). Handbook of social choice and welfare. North Holland: *Gulf Professional Publishing*.

[14] Moulin, H. (1998). Axioms of Cooperative Decision Making. Cambridge: *Cambridge University Press*.

[15] Fishburn, P.C. (1968). Theory of social choice. Princeton: *Princeton University Press*.

[16] Zhuravlev, I. Yu. & Gurevitch, I. B. (1989). Pattern recognition and image recognition. In Pattern recognition, classification, forecasting: Mathematical techniques and their application; Zhuravlev, I. Yu. Ed. *Nauka: Moscow.*, *2*, 5-72. (in Russia)

[17] Yurin, A. Y. (2012). An approach for definition of recommendations for prevention of repeated failures with the aid of case-based reasoning and group decision-making methods. *Expert Syst Appl.*, *39*, 9282-9287.

[18] Nikolaychuk, O. A. & Yurin, A. Y. (2008). Computer-aided identification of mechanical system's technical state with the aid of case-based reasoning. *Expert Syst Appl.*, *34*, 635-642.

[19] Berman, A. F. & Nikolaychuk, O. A. (2007). Technical state space of unique mechanical systems. *Journal of Machinery Manufacture and Reliability.*, *36*(1), 10-16.

[20] Berman, A. F., Nikolaychuk, O. A., Pavlov A. I. & Yurin A. Yu. (2010). Provision of Safety for Technological Systems with the Aid of Case-Based Reasoning. Case-Based Reasoning: Processes, Suitability and Applications (Editors: Antonia M. Leeland). NY. Nova Science Publishers.

In: Case-Based Reasoning
Editor: Tao Lin

ISBN: 978-1-63482-952-6

Chapter 3

SEMANTIC SIMILARITY

Rafael Batres [*]
Tecnológico de Monterrey, Campus Cuernavaca, México

ABSTRACT

Semantic similarity measures are concerned with quantification of the likeness of the meanings of features that are compared. When features are obtained from classes in a taxonomy, semantic similarity is a function that assigns a numerical value to the similarity between two classes of objects. For example, two classes that have a direct common superclass are semantically more similar than two classes that are far apart in the taxonomy. CBR systems can use semantic approaches to retrieve cases. Typically, such CBR systems utilize a knowledge representation method combined with an approach for measuring the similarity of the meaning of things represented in the cases. The latter is known as semantic similarity. This chapter describes traditional similarity measures as well as recent approaches and applications.

Keywords: semantic similarity, ontologies, information content, path-based measures

INTRODUCTION

CBR systems can use semantic approaches to retrieve cases, enabling candidate cases similar to a new problem to be identified so that their solutions can be used. As in traditional CBR systems, cases in the case base with features similar to the features of the new problem are recalled. However, in addition, semantic features can be defined in terms of classes of objects or instances. Typically, such CBR systems utilize a knowledge representation method combined with an approach to measure the similarity of the meaning of the things represented in the cases. The latter is known as semantic similarity. This chapter describes traditional similarity measures, as well as recent approaches and applications.

[*] Autopista del Sol Km. 104, Xochitepec 62790, Mexico. Email: rafael.batres@itesm.mx

Semantic similarity measures are concerned with quantifying the likeness of the meanings of the features compared. When features are obtained from classes in a taxonomy, a semantic similarity is a function that assigns a numeric value to the similarity between two classes of objects. For example, two classes that have a direct common superclass are semantically more similar than two classes that are far apart in the taxonomy.

The classes in a taxonomy are related by means of a subclass relation, also known as *is-a* relation or subsumption relation. A class *B* is said to be a subclass of *A* if all the members of *B* are also members of *A*. Therefore, strictly speaking, the similarity between two classes refers to the comparison of generic members of those classes.

A class represents a category of things that share a set of attributes. For example, a class *person* has attributes such as place of birth and date of birth.

An instance is a member of a class. For example, Mark Zuckerberg is an instance of class *person*. In CBR systems that use semantic similarities, cases refer to particular instances, but it is still useful to compare the classes to which these instances belong.

Class taxonomies are defined using the subclass relation. A class is a subclass of another class if every instance of the subclass is also an instance of the superclass. Figure 1 shows an example of a taxonomy. In this taxonomy, *B* and *D* are subclasses of *A*, *C* and *E* are subclasses of *B*, *F* is a subclass of *D*, and *G* is a subclass of *E*.

The subclass relation is transitive, which means that if *B* is a subclass of *A* and *C* is a subclass of *B*, then *C* is also a subclass of *A*. This implies that inheritance of attributes occurs between classes linked through the subclass relation. For example, according to the taxonomy in Figure 2, the attribute *maximum_payload_capacity*, which is owned by the class *vehicle*, is inherited down to *taxi cab*.

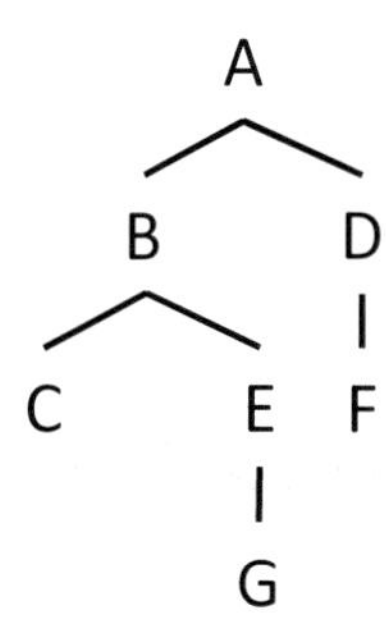

Figure 1. A sample taxonomy. A to G are classes.

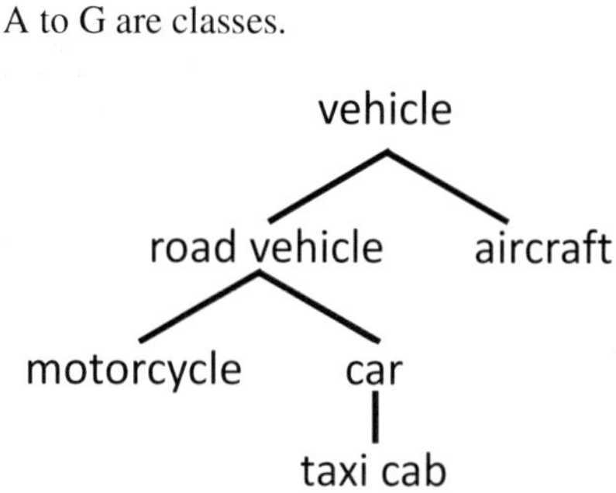

Figure 2. A small vehicle taxonomy.

Instances can have relations other than the subclass relation. In general, a relation is a function that maps its arguments to a Boolean value of True. Commonly used relations include mereological, topological, and causal relations.

A mereological relation expresses that something is a part of a whole. A mereological relation is reflexive, which means that everything should be part of itself (x is a part of x). Mereological relations are also characterized by the fact that distinct things cannot be part of each other (if x is a part of y, and y is a part of x then $x=y$). Mereological relations are transitive, meaning that any part of any part of a thing is itself part of that thing (if x is a part of y and y is a part of z then x is a part of z).

Topology refers to the connectivity between objects (Clarke, 1981). A topological relation is defined as symmetric and transitive. This implies that if x is connected to y, then y is also connected to x. Moreover, if x is connected to y and y is connected to z, then x is also connected to z.

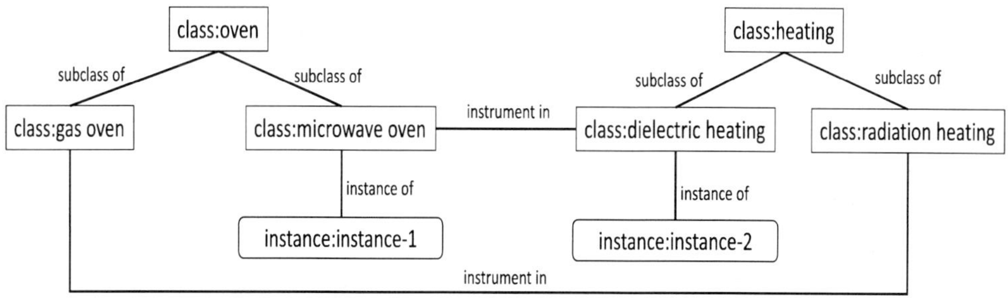

Figure 3. Instances of two classes of ovens.

In case-based reasoning, it is crucial to take into account the relations that exist between physical objects and processes. For example, instance-1 and instance-2 in Figure 3 are similar because both are intended to generate heat when used. This is despite the fact that radiation is realized in the former and dielectric heating in the latter. Furthermore, both appliances use electricity and convert it to heat. A semantic similarity is a function that assigns a numeric value to the similarity between two classes of objects. The classes in a taxonomy are related by means of a subclass relation, also known as *is-a* relation or subsumption relation. A class C_1 is said to be a subclass of C_2 (C_2 is a superclass of C_1) if all the members of C_1 are also members of C_2. The similarity between two classes is based on how closely they are related in the taxonomy.

The following are definitions that will be used throughout the remainder of this chapter.

Least common ancestor (lca(c_i, c_j)). Least common ancestor of c_i and c_j. The most specific classes that both classes have in common.

Shortest path (minpath(c_i,c_j)). The length of the shortest path from class c_i to class c_j in the taxonomy.

Depth (depth(c_i)). The length of the shortest path from class c_i to the root entity.

Maximum depth. The maximum depth(c_i) of the taxonomy.

Taxonomy-based semantic similarity measures are typically classified into three groups: path-based, information content, and attribute-based measures.

Path-Based Measures

Path-based measures are based on the distance between two classes. Typically, such distance is calculated as a function of the number of edges or nodes on the shortest path linking the two classes. When the relations between classes are exclusively taxonomic relations, the distance between two classes is equal to the sum of the two shortest paths from the classes to their closest common ancestor. The shorter the path, the more similar the classes. For example, when comparing *G* and *D* in Figure 1, the closest common ancestor is *B*, so the length of the shortest path between these classes is 3. Likewise, the path length between *D* and *F* is 4.

The simplest path-based similarity measure is the length of the shortest path linking the two classes added to 1.

$$sim_{Path}(C_i, C_j) = 1 + \text{minpath}(C_i, C_j)$$

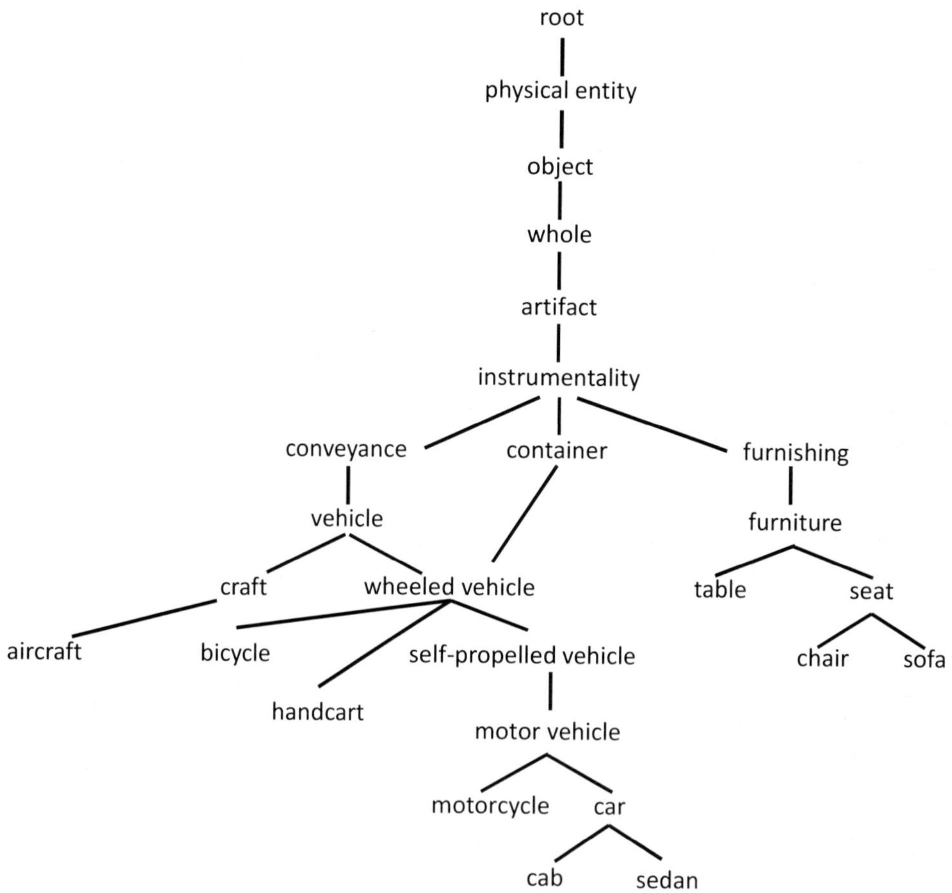

Figure 4. A fragment of the WordNet taxonomy.

Problem 1. Figure 4 shows a fragment of the taxonomy used in the WordNet lexical database. Calculate the similarity between *car* and *aircraft*.

Solution. The least common ancestor is *vehicle*. The length of the minimum path between *car* and *aircraft* is thus the length of the path between *car* and *vehicle* (car>motor vehicle>self-propelled vehicle>wheeled vehicle>vehicle) added to the length of the path between *vehicle* and *aircraft* (aircraft>craft>vehicle). The number of edges of this path is 6. Thus,

$$sim_{Path}(\text{car}, \text{aircraft}) = 1 + \text{minpath}(\text{car}, \text{aircraft}) = 1 + 6 = 7$$

The similarity measure of Leacock and Chodorow (Leacock et al., 1998) normalizes the shortest path between two classes by twice the maximum depth of the taxonomy.

$$sim_{LCH}(C_i, C_j) = -\log \frac{\text{minpath}(C_i, C_j)}{2D_{\max}}$$

where D_{max} is the maximum depth of the taxonomy, defined as the length of the longest path from a class to the root class of the taxonomy.

Problem 2: calculate the similarity measure of Leacock and Chodorow between classes *G* and *D* in Figure 1.

Solution. The length of the shortest path between the two classes is minpath(G,D)=3. The longest path between a class and the root of the taxonomy is the path between *G* and *A*, where *A* is the root class. As a result, the maximum depth is $D_{max} = 3$ and $sim_{LCH}(G, D) = -\log\left(\frac{3}{6}\right) = 0.30$.

As the distance of a class to itself is 0, the equation of Leacock and Chodorow is often modified to the following expression.

$$sim_{LCH}(C_i, C_j) = -\log \frac{\text{minpath}(C_i, C_j) + 1}{2D_{\max} + 1}$$

Therefore, the values of sim_{LCH} are between 0 and $\log(2D_{max} + 1)$.

Problem 3. Repeat Problem 1, using the similarity measure of Leacock and Chodorow between *car* and *aircraft*. The maximum depth is 20.

Solution:

$$sim_{LCH}(\text{car}, \text{aircraft}) = -\log \frac{6+1}{2(20)+1} = -\log \frac{7}{41} = 0.76$$

Problem 4. Using the taxonomy of Figure 4, calculate the similarity between *car* and *furnishing*.

Solution. The least common ancestor is *instrumentality*. In this case, the shortest path is car>motor vehicle>self-propelled vehicle>wheeled vehicle>container>instrumentality + furnishing>instrumentality. The length of this path is 6.

Resnik (1995) defined a similarity measure based on the amount of information shared by two classes. The amount of shared information is determined by the information content of the least common ancestor of the two classes.

$$sim_{\text{Resnick}}(C_i, C_j) = -\log p(\text{lca}(C_i, C_j)) = \text{IC}(\text{lca}(C_i, C_j))$$

Though the Resnik measure is simple, it may not be able to distinguish between closely similar classes, since many classes may share the same least common ancestor and, as a result, have identical values of similarity (Pedersen, 2007).

Problem 7. Use the Resnik equation to calculate the similarity between *sedan* and *bicycle*. The information content values are shown in parenthesis in Figure 5 and were calculated using the SemCor corpus[1]. Compare against the similarity between sedan and handcart.

Solution. The least common ancestor is wheeled vehicle. Thus, $\text{IC(wheeled vehicle)} = 6.31$. This value is exactly the same as the value obtained with the Resnik equation for the pair (sedan, handcart), because in both cases the least common ancestor is *wheeled vehicle*.

Jiang and Conrath (1997) extend Resnik's similarity by combining the information contents of C_i, C_j and their lowest common ancestor.

$$sim_{\text{JC}}(C_i, C_j) = \frac{1}{\text{IC}(C_i) + \text{IC}(C_j) - 2\,\text{IC}(\text{lca}(C_i, C_j))}$$

Another extension of Resnik's similarity is that of Lin, in which the Resnik similarity measure appears in the nominator (1998). Lin's similarity measure is defined as

$$sim_{\text{Lin}}(C_i, C_j) = \frac{2\,\text{IC}(\text{lca}(C_i, C_j))}{\text{IC}(C_i) + \text{IC}(C_j)}$$

The underlying assumption of Resnik similarity is that the more information content two classes share, the more similar they are. On the other hand, in addition to Resnik's assumption, Jin-Conrath and Lin extensions assume that the less information content two classes share, the less similar they are.

Problem 8. Repeat Problem 7, using the Lin extension. The necessary values of information content are: $\text{IC(wheeled vehicle)} = 6.31$, $\text{IC(sedan)} = 11.072$, $\text{IC(bicycle)} = 10.156$ and $\text{IC(handcart)} = 10.667$.

Solution.

$$sim_{\text{Lin}}(\text{sedan}, \text{bicycle}) = \frac{2(6.31)}{11.072 + 10.156} = 0.594$$

$$sim_{\text{Lin}}(\text{sedan}, \text{handcart}) = \frac{2(6.31)}{11.072 + 10.667} = 0.580$$

[1] The SemCor corpus contains about 360,000 words of American English which have been tagged with syntactic and semantic information as part of the WordNet project.

An alternative information content function proposed by Seco et al. (2004) uses the taxonomy structure without the corpus. The value of $IC(C_k)$ is obtained by counting the number of all subclasses of C_k, and then dividing that value by the total number of classes in the taxonomy.

$$IC(C_k) = 1 - \frac{\log(h(C_k)+1)}{\log(S)}$$

where $h(C_k)$ is the number of subclasses of C_k and S is the total number of classes in the taxonomy. This approach requires the taxonomy to be as complete as possible. In other words, the taxonomy should include most of the specializations of a specific class in order to provide reliable results. As a result, partial taxonomies with a limited scope may not be suitable for this purpose (Sánchez et al., 2012).

The advantage of this approach is that no corpus is required if a taxonomy is available. However, classes having the same number of subclasses but different degrees of generality have the same IC value. Zhou et al. (2008) overcome this problem by including the relative depth of the class:

$$IC(C_k) = w\left(1 - \frac{\log(h(C_k)+1)}{\log(S)}\right) + (1-w)\left(\frac{\log(\mathrm{depth}(C_k))}{\log(D_{\max})}\right)$$

where w is a weight with a typical value of $w = 0.5$.

Attribute-Based Measures

Attribute-based similarity measures are functions of properties or relations to other classes in the taxonomy. Each class is described in terms of a set of properties or relations. Common attributes tend to increase the similarity of the classes being compared. Conversely, uncommon features tend to diminish the similarity.

Attribute-based similarities have their origin in the work of Tversky (1977) whose similarity measure is based on set theory. Tversky's similarity measure is defined as

$$\mathrm{sim}_{Tversky}(A_i, A_j) = \frac{|A_i \cap A_j|}{|A_i \cap A_j| + \alpha|A_i \setminus A_j| + \beta|A_j \setminus A_i|}$$

where A_i and A_j are sets of attributes of classes C_i, C_j respectively, and α and β are parameters. $|A_i \setminus A_j|$ is the set of features in A_i but not in A_j, and $|A_j \setminus A_i|$ is the set of features in A_j not in A_i. Literature regarding the use of attribute-based similarities is centered on the use of synsets for the comparison between words. However, attributes of ontology classes can also be used.

Rodriguez and Egenhofer (2003) defined and α and β as:

$$\alpha = \begin{cases} \dfrac{\text{depth}(C_1)}{\text{depth}(C_1) + \text{depth}(C_2)} & \text{if depth}(C_1) \leq \text{depth}(C_2) \\ 1 - \dfrac{\text{depth}(C_1)}{\text{depth}(C_1) + \text{depth}(C_2)} & \text{if depth}(C_1) > \text{depth}(C_2) \end{cases}$$

$$\beta = 1 - \alpha$$

When $\alpha = 0.5$ and $\beta = 0.5$, Tversky's equation becomes the Dice's coefficient, which quantifies the overlap of two sets in relation to an estimate of their average size (Dice, 1945).

$$\text{sim}_{Dice}(A_i, A_j) = \frac{|A_i \cap A_j|}{\frac{1}{2}(|A_i| + |A_j|)}$$

When $\alpha = \beta = 1.0$, equation (4) becomes the Jaccard's coefficient in which the denominator is the union of the cardinalities of the sets.

$$\text{sim}_{Jaccard}(A_i, A_j) = \frac{|A_i \cap A_j|}{|A_i \cap A_j| + |A_i \setminus A_j| + |A_j \setminus A_i|} = \frac{|A_i \cap A_j|}{|A_i \cup A_j|}$$

The Jaccard's coefficient is related to the Dice's similarity through the following equation:

$$\text{sim}_{Jaccard}(A_i, A_j) = \text{sim}_{Dice}(A_i, A_j)/(2 - \text{sim}_{Dice}(A_i, A_j))$$

A variation of the Dice equation is the overlap coefficient (Han and Kamber, 2006):

$$\text{sim}_{Overlap}(A_i, A_j) = \frac{|A_i \cap A_j|}{\min(|A_i|, |A_j|)}$$

Another variation is the all-confidence similarity:

$$\text{sim}_{All-confidence}(A_i, A_j) = \frac{|A_i \cap A_j|}{\max(|A_i|, |A_j|)}$$

The cosine similarity can also be considered attribute-based if sets are used instead of vectors.

$$\text{sim}_{\text{Cosine}}(A_i, A_j) = \frac{|A_i \cap A_j|}{\sqrt{|A_i|}\sqrt{|A_j|}}$$

Problem. Electric kettles are characterized by the following attributes: consumes electricity, heats water, and produces hot water. Microwave ovens are characterized by: consumes electricity, generates high-energy-frequency waves, heats, and heats food. Use the overlap similarity measure to calculate the similarity between these two electric appliances.

Solution.

A_i = { consumes electricity, heats water, and produces hot water }

A_j = { consumes electricity, generates high-energy-frequency waves, heats, and heats food }

$$|A_i| = 3$$

$$|A_j| = 4$$

$$A_i \cap A_j = \{\text{consumes electricity, heats}\}$$

$$|A_i \cap A_j| = 2$$

$$\text{sim}_{Overlap}(A_i, A_j) = \frac{|A_i \cap A_j|}{\min(|A_i|, |A_j|)} = \frac{2}{\min(3,4)} = \frac{2}{3} = 0.666$$

Semantic Sources

Semantic similarity measures rely on the use of semantic sources that describe classes of things. Semantic sources define the classes in terms of taxonomies and other knowledge-based elements such as attributes. Some practical semantic sources are WordNet, ontologies and Formal Concept Analysis (FCA), which are described below.

WordNet

WordNet® is a lexical database of English words, which is publicly available (Princeton University, 2010). Words are grouped into sets of cognitive synonyms called synsets, each expressing a distinct concept. Additionally, a synset contains a gloss, which is a brief textual definition.

Words are linked by means of *is-a* and other semantic relations. The *is-a* relation, also known as hyperonymy or hyponymy, is the building block of the WordNet taxonomy. In addition, WordNet implements a mereological relation known as meronymy. For example, a chair is described in terms of its parts, which are *backrest*, *seat* and *leg*.

WordNet not only defines classes of things (such as person or country) but also instances (such as specific persons or countries).

Information-content measures based on WordNet also require a corpus, a large and structured collection of texts. One corpus that is widely used is SemCor (Miller et al., 1993), which has been tagged with syntactic and semantic information based on WordNet.

Some software resources for calculating semantic similarities based on WordNet are:

1. WordNet::Similarity – http://wn-similarity.sourceforge.net
2. Natural Language Toolkit (NLTK) – http://nltk.org
3. WordNet Similarity for Java (ws4j) – https://code.google.com/p/ws4j/
4. DKPro Similarity – https://code.google.com/p/dkpro-similarity-asl/

Ontologies

Ontologies are models that use mathematical logic to define classes of things and their relations (Batres and Akmal, 2013). A class represents a set of things that share the same attributes. For example, an instance of nucleus is (or has been) a part of a cell.

A relation is a tuple that represents a relationship between two or more things. Examples of relations are *connected-to*, and *part-of*. The subclass or *is-a* relation is a type of relation for describing the position of a class in regard to others in the taxonomy. A class is a subclass of another class if every member of the subclass is also a member (instance) of the superclass. For example, an instance of a class *chair* is also an instance of a class *furniture*.

Axioms are typically represented as logic constructions that formally define a given class or relation. For example, the *connected-to* relation is defined in such a way that the computer can conclude that *B* is connected to *A* provided that *A* is connected to *B*. Another example is the *part-of* relation which is defined with the axiom "if x is a part of y and y is a part of z then x is a part of z."

One of the benefits of ontologies is that they can be processed by knowledge reasoning algorithms so that hidden relations can be discovered. A variety of tools are publicly available for reasoning, editing, and syntax checking.

A new ontology can extend other ontologies, avoiding reinventing the wheel. Some existing ontologies are: the ISO 15926 (Batres et al., 2007), SUMO (Pease et al., 2002) and OntoCAPE (Marquardt et al., 2010). A very convenient tool for ontology creation is the Protégé ontology editor (http://protege.stanford.edu/).

Formal Concept Analysis (FCA)

Formal Concept Analysis (FCA) is a technique for knowledge processing based on applied lattice and order theory (Priss, 2006; Wille, 1982). FCA can be used to generate taxonomies, and is also convenient as a source of attribute information that can be used in attribute-based measures (Batres and Akmal, 2013). FCA is based on the so-called formal context, an incidence matrix that contains objects and attributes. Objects are listed in the rows and the attributes in the columns of the table. If an object has an attribute, which means that there is a binary relation between them, a '1' (equivalent to a checkmark) is inserted in the corresponding cell, otherwise the cell has the value of '0' (equivalent to an empty cell with no checkmark).

A formal context is processed to identify the so-called formal concepts, which are sets of objects that share the same attributes. In addition, formal concepts can be partially ordered into a lattice, such that a concept is a subconcept of another concept. An edge in the lattice means that a concept is a subconcept of another concept. The relation attached to an edge is transitive, so that if a node A is a subconcept of B, and B is also a subconcept of C, A is a subconcept of C. As result, a subconcept inherits all the attributes from all its superconcepts. Therefore, if potential classes are used as objects in the context table, the resulting lattice represents a class taxonomy.

Several lattice-construction algorithms have been proposed in which some concepts are obtained by the intersection of others. Once the lattice has been generated, it can be visualized and analyzed.

Several software tools are available that implement the FCA methodology and algorithms. One such tool is Concept Explorer, which is freely available (http://conexp.sourceforge.net/users/).

REFERENCES

Batres, R., and S. Akmal, A Formal Concept Analysis-Based Method for Developing Process Ontologies, *Journal of Chemical Engineering of Japan*, Vol. 46, No. 6, pp. 396-406 (2013).

Batres, R., M. West, D. Leal, D. Price, K. Masaki, Y. Shimada, T. Fuchino and Y. Naka, "An Upper Ontology based on ISO 15926", *Computers And Chemical Engineering*, vol. 31, p. 519-534, 2007.

Clarke, B. L., A calculus of individuals based on 'Connection', *Notre Dame Journal of Formal Logic*, Vol. 22, pp. 204-218 (1981).

Dice, L. R., Measures of the Amount of Ecologic Association Between Species. *Ecology, Vol.* 26, No. 3, pp. 297-302 (1945).

Han, J. and Kamber, M., Data Mining: Concepts and Techniques, 2nd ed. Morgan Kaufmann Pub., an imprint of Elsevier (2006).

Jiang, J., Conrath, D., Semantic similarity based on corpus statistics and lexical taxonomy, In Proceedings of the international conference on research in computational linguistics (ROCLING X), Taiwan, pp. 19-33 (1997).

Leacock, C., Chodorow, M., Miller, G. A., Combining local context and WordNet similarity for word sense identification, In: WordNet: an electronic lexical database. MIT Press, pp. 265-283 (1998).

Lin, D., An information-theoretic definition of similarity, In: Proceedings of the 15th international conference on machine learning (ICML98). Madison, Wisconsin, USA, pp. 296-304 (1998).

and energy, OAPID can be used to analyze the participation level of variables in each oscillation mode, i.e., eigenvectors and participation factors comprehensively.

The small signal stability model of power system can be expressed as:

$$\begin{cases} \dot{X} = AX + BU \\ Y = CX \end{cases} \tag{3}$$

Where X is $n\times1$ variable vector, U is $n\times1$ input vector, Y is $n\times1$ output vector; A is $n\times n$ coefficient matrix, B is $n\times q$ input matrix, C is $m\times n$ output matrix.

In actual application, elements of output vector Y are active powers of generators, lines and loads. Provided ΔP_i and ΔP_j are two oscillatory active power increments, which can be expressed as formula (4) and (5):

$$\Delta P_i = C_{i1}x_1 + C_{i2}x_2 + \cdots + C_{in}x_n \tag{4}$$

$$\Delta P_j = C_{j1}x_1 + C_{j2}x_2 + \cdots + C_{jn}x_n \tag{5}$$

Ratio of ΔP_i and ΔP_j is:

$$\frac{\Delta P_i}{\Delta P_j} = \frac{C_{i1} + C_{i2}\, x_2/x_1 + \cdots + C_{in}\, x_n/x_1}{C_{j1} + C_{j2}\, x_2/x_1 + \cdots + C_{jn}\, x_n/x_1} \tag{6}$$

For a certain oscillation mode (eigenvalue) λ, ratio of two state variables is equal to the ratio of elements in corresponding right eigenvector. Provided the right eigenvector of λ is v, and its elements are $v1, v2, \cdots, vn$, formula (6) can be written as:

$$\frac{\Delta P_i}{\Delta P_j} = \frac{C_{i1} + C_{i2}\, v_2/v_1 + \cdots + C_{in}\, v_n/v_1}{C_{j1} + C_{j2}\, v_2/v_1 + \cdots + C_{jn}\, v_n/v_1} \tag{7}$$

for ratio of multiple increments of active powers:

$$\begin{aligned} \Delta P_1 : \Delta P_2 : \cdots : \Delta P_m = (\sum_{k=1}^{n} C_{1k}\, v_k/v_1) : \\ (\sum_{k=1}^{n} C_{2k}\, v_k/v_1) : \cdots : (\sum_{k=1}^{n} C_{mk}\, v_k/v_1) \end{aligned} \tag{8}$$

Because it is difficult to conduct eigenvalue analysis online for large scale power systems, above OAPID index is not available. Literature [14] points out that OAPID index can be obtained using WAMS data via Prony algorithm. TLS-ESPRIT algorithm [15] has better anti-noise performance and higher accuracy than Prony algorithm, thus OAPID index of interested oscillation mode can be obtained directly using WAMS or time domain simulation data, via TLS-ESPRIT algorithm instead of Prony algorithm. Further, normalization is carried out according to the maximum oscillatory active power increment.

SMALL SIGNAL STABILITY EARLY WARNING BASED ON CBR

Provided structure and parameters of a regional power gird are certain, its small signal stability mainly relates to operation status. As low frequency oscillations occurring in large scale power grid usually appear with certain rules and characteristics, dispatchers can take similar measures to mitigate the power oscillations. Therefore, small signal stability early warning function for a regional power grid is proposed in this chapter based on CBR theory. The historical case base is built up off line according to operation status, results of eigenvalue analysis, as well as oscillation information recorded by WAMS. And then, small signal stability on line warning under current operation status is achieved by case matching via CBR theory. In detail:

a) Building up historical case base

Based on given power system model and operation status recorded by Energy Management System (EMS), small signal stability under typical operation status is obtained via eigenvalue analysis or OAPID theory. According to formula (1), information characterizing each operation status and resulting small signal stability are treated as condition attribute and decision attribute of a historical case, respectively. Historical case base is built up based on these historical cases, as case representation and organization in CBR theory. Depending on whether low frequency oscillation occurs, the historical case base is classified into two types, namely base 1 (no low frequency oscillation occurs) and base 2 (low frequency oscillations occur). The contents of base 1 and base 2 are shown in table 1.

Table 1. Contents of case bases

	Base 1 (no low frequency oscillation occurs)	Base 2 (actual or simulated low frequency oscillations occur)
Condition attribute	Information of operation status	Information of operation status
Decision attribute	corresponding results of eigenvalues analysis	corresponding results of eigenvalues analysis or OAPID theory

b) Case matching via CBR theory

To search the case with highest similarity with current operation status based on EMS and WAMS realtime data. Depending on whether low frequency oscillation occurs or not, forward matching based on condition attributes and backward matching based on decision attributes are carried out:

b-1) no low frequency oscillation

If there is no power oscillation detected by WAMS, forward matching is carried out in base 1 by matching current operation status with the condition attribute of each historical case in base 1. And the decision attribute of the matched (most similar) case is treated approximately as evaluation result of small signal stability under current operation status. If damping ratio for any inter-area oscillation mode in decision attribute of the matched case is lower than setting (usually 5%), early warning of small signal stability is released. Further, backward matching is carried out in base 1 by matching both decision and condition attributes. A historical case in base 1, which has sufficient damping ratio for any inter-area oscillation mode and the highest similarity to current operation status, is used as advice to dispatchers for power system re-schedule.

b-2) low frequency oscillation

If there is power oscillation detected by WAMS, forward matching is carried out in base 2 by matching current operation status with the condition attribute of each historical case in base 2, as well as matching current oscillation frequency and OAPID index with those in decision attribute of each historical case in base 2. Based on oscillation frequency and OAPID index of the oscillation mode in decision attribute of the matched case, backward matching is carried out in base 1 by matching both decision and condition attributes. A historical case in base 1, which has sufficient damping ratio for the oscillation mode and the highest similarity to current operation status, is used as advice to dispatchers to adjust operation status.

CASE STUDIES

Case studies are carried out using EMS and WAMS data in a regional power grid (with 78 PMUs), to verify effectiveness of the proposed small signal stability early warning function. In these case studies, historical case base 1 consists of 46 cases, and twelve of them are with weak damping ratio; historical case base 2 consists of 7 cases of inter-area low frequency power oscillations. Damping ratio threshold is 5% in the case studies.

Normal Operation Status

Based on information of a normal operation status of regional power grid, which is the condition attribute used by forward matching in base 1, decision attribute of the matched case is obtained, and information of the corresponding inter-area low frequency power oscillation mode with lowest damping ratio are listed in tables 2 and 3, respectively.

Table 2.The matched case and corresponding oscillation mode

Case/mode	Case similarity	Real part of eigenvalue	Imaginary part of eigenvalue	Frequency/Hz	Damping ratio/%
11	0.9378	-0.2994	3.2371	0.5152	9.21

Table 3. Correlative generators and participation factors

Correlative generators	participation factors	Correlative generators	participation factors
JHF-G1	0.0140	MHS-G6	0.0066
JLD-G2	0.0082	MHX-G1	0.0061
MLD-G2	0.0081	MHX-G2	0.0061
MLD-G1	0.0081	JLD-G1	0.0057
JYE-G2	0.0066	JSE-G1	0.0055

It can be seen in table 2 that the lowest damping ratio of oscillation mode in the matched case obtained by forward matching is larger than the threshold. Thus, the power grid under current operation status has sufficient damping, i.e., small signal stability.

Operation Status with Low Frequency Power Oscillation Occurrence

Active power oscillation occurred and was detected by WAMS in the regional power grid, and its oscillation frequency is 0.808Hz. Forward matching is carried out in case base 2 according to operation status, oscillation frequency and OAPID index. The matched case/mode in case base 2 has relatively high similarity (0.831249), similar oscillation frequency and OAPID index, which are listed in table 4.

Table 4. Comparison of OAPID index obtained by TLS-ESPRIT with that of the matched case in base 2

Generators/ lines	OAPID obtained by TLS-ESPRIT		Generators / lines	OAPID of the matched case/mode	
	Relative amplitude	Relative phase		Relative amplitude	Relative phase
DBS-TP1	1	0	DBS-TP1	1	0
YSH-G3	0.439	-46.689	YSH-G3	0.597	72.671
YSH-G4	0.397	-51.736	YSH-G4	0.547	80.288
YYM-G5	0.338	-28.278	YYM-G5	0.373	5.200
YQB-G2	0.263	-41.620	YQB-G2	0.493	11.547
YMS-G1	0.261	-45.445	YMS-G1	0.273	37.034
KH-XF1	0.239	-14.317	KH-XF1	0.483	29.506
YHB-G5	0.213	61.859	YHB-G5	0.478	8.555
YQB-G3	0.192	67.962	YQB-G3	0.548	30.631
YYH-G3	0.191	-35.562	YQB-G4	0.547	30.759

It can be found from table 4 that percentage of same generators and lines, which are with relatively large OAPID index, is 90%. Further, provided relative phase of a generator (or line) with the maximum amplitude is 0, generators and lines with relative phase within -45°~ 45°or

135°~ 225°are shown in table 5. It can be found from table 5 that percentage of same generators and lines, which are in the same range of relative phase, is 75%.

Table 5. Comparison of the relative phase of OAPID index obtained by TLS-ESPRIT with that of the matched case in base 2

Phase range	OAPID obtained by TLS-ESPRIT	OAPID of the matched case/mode
- 45°~ 45°	YYM-G5, YQB-G2 KH-XF1 line	YYM-G5, YQB-G2, YMS-G1, KH-XF1 line, YHB-G5, YQB-G3
135°~ 225°	none	none

As the outputs of forward matching, results of eigenvalue analysis of the matched case are shown in table 6 and table 7, respectively.

Table 6. Oscillation mode information of the matched case in base 2

Case/mode	Case similarity	Real part of eigenvalue	Imaginary part of eigenvalue	Frequency/Hz	Damping ratio/%
1	0.831249	-0.073143	4.355848	0.693255	1.68

Table 7.Generators with high participation factors to the oscillation mode of the matched case

Correlative generators	participation factors	Correlative generators	participation factors
YSY-G3	0.4898	YQB-G1	0.0004
YSY-G1	0.0020	YMQ-G1	0.0004
YSY-G2	0.0020	YQB-G2	0.0003
YDBS-G1	0.0009	YHB-G6	0.0003
YDBS-G2	0.0008	YQB-G4	0.0003

Obviously, the oscillation mode information obtained by forward matching is in line with actual situation recorded by WAMS, and OAPID index during low frequency oscillation is also in line with the result of eigenvalue analysis.

To verify the correctness and effectiveness of forward matching, eigenvalue analysis is also carried out based on steady state operation status of the power grid recorded by EMS [14-16] before the occurrence of low frequency oscillation. The results are listed in tables 8 and 9, respectively.

Table 8. Oscillation mode information of actual operation status

Real part of eigenvalue	Imaginary part of eigenvalue	Frequency/Hz	Damping ratio/%
-0.0646	4.6168	0.7348	1.3988

Comparing the results in tables 6-9, it can be found that oscillation frequency, damping ratio, sorting of correlative generators and participation factors are very similar, which illustrates that forward matching is correct and the results are in line with actual situation.

To make efficient adjustment of operation status, backward matching is carried out in case base 1, and results of eigenvalue analysis of the matched case are shown in tables 10 and 11, respectively.

Table 9. Correlative generators and participation factors of actual operation status

Correlative generators	participation factors	Correlative generators	participation factors
YSY-G3	0.4846	YLY-G1	0.0003
YSY-G2	0.0046	YQB-G3	0.0003
YSY-G1	0.0035	YQB-G4	0.0003
YDBS-G1	0.0007	YQB-G2	0.0002
YDBS-G2	0.0007	YKH-G7	0.0002

Table 10.Oscillation mode information of the matched case in base 1

Case/mode	Case similarity	Real part of eigenvalue	Imaginary part of eigenvalue	Frequency/Hz	Damping ratio/%
40	0.93603	-0.336298	5.575946	0.887439	6.0203

Table 11. Generators with high participation factors to the oscillation mode of the matched case

Correlative generators	participation factors	Correlative generators	participation factors
YSY-G3	0.3391	YSH-G3	0.0015
YSY-G2	0.0922	YSH-G4	0.0015
YSY-G1	0.0907	YQB-G3	0.0012
YDBS-G1	0.0031	YQB-G4	0.0012
YDBS-G2	0.0025	YLY-G1	0.0010

Table 12. Operation status adjustment obtained by backward matching

Node	Trans-mission line	Similarity	Actual active power (MW)	Actual reactive power (Mvar)	Actual voltage (kV)	Suggested active power (MW)	Suggested reactive power (Mvar)	Suggested Voltage (kV)
4	SH-SZ1	0.89407	281.219	-59.3527	534.786	272.431	145.65	534.054
7	YM-ZZ	0.94279	460. 879	45.0442	543.592	515.376	-141.01	531.3013
19	ML-DGH1	0.95305	-137.331	-59.3271	535.695	-141.823	45.15	534.5609
11	SH-YX1	0.96252	409.793	-18.6769	535.124	392.479	-159.021	529.5052
16	YX-XL1	0.97239	-288.5	-60.5	535.317	-286.861	-184.527	535.3411
17	ZJB-ES	0.97338	65	0	534.294	70.423	117.225	531.539
12	SXZY-LQ1	0.98714	1087.5	83.5	536.597	1176.127	-10.398	531.6036
21	DJ-CH1	0.98745	142.55	-23.8957	533.578	197.824	44.041	535.4831

Obviously, the oscillation mode information obtained by backward matching are very close to those in table 9, in view of oscillation frequency, sorting of correlative generators and participation factors. Further, damping ratio of the oscillation mode obtained by backward matching is improved greatly than that in table 8. As the output of backward matching, the adjusted operation status (including 8 power plants and substations) is listed in table 12.

Table 13. Results of eigenvalue analysis of the suggested operation status

Real part of eigenvalue	Imaginary part of eigenvalue	Frequency/Hz	Damping ratio/%
-0.2972	5.4466	0.8669	5.4481%

Results of eigenvalue analysis of the adjusted operation status are shown in table 13, which illustrate the new operation status has much higher damping ratio.

Conclusion

In this chapter, a method based on CBR theory and OAPID index is proposed to improve small signal stability of a regional power grid. This method makes full use of control strategies and experiences of the power grid to achieve the early warning and suppression of inter-area low frequency oscillations. Further, the proposed method provides appropriate instructions to dispatchers to reschedule operation status of the power grid. Case studies demonstrate the effectiveness of the method.

References

[1] Y.V. Makarov, V.A. Maslennikov, and D.J. Hill, 1996. Revealing load shaving the biggest influence on power system small disturbance stability. *IEEE Trans. Power Systems*, 11, 2018-2023.

[2] J.V. Milanovic, I.A. Hiskens, 1995. Effects of load dynamics on power system damping. *IEEE Trans. Power Systems*, 10, 1022-1028.

[3] J. Daniel, Tylavsky, Anjan Bose, 1992. Parallel Processing in Power System Computation. *IEEE Trans. Power System*, 7, 629-638.

[4] N. Kakimoto, M. Sugumi, T. Makino, and K. Tomiyama, 2006. Monitoring of interarea oscillation mode by synchronized phasor. *IEEE Trans. Power Systems*, 21, 260-268.

[5] Kakimoto, M. Sugumi, T. Makino, and K. Tomiyama, 2006. Monitoring of interarea oscillation mode by synchronized phasor measurement. *IEEE Trans. Power Systems*, 21, 260– 268.

[6] A.Z. Gamm, Yu.A. Grishin, I.N. Kolosok, A.M. Glazunova, E.S. Korkina, 2007. New EPS state estimation algorithms based on the technique of test equations and PMU measurements. *Power Tech*. 2007 IEEE, Lausanne, 1670– 1675.

[7] Borka Milosevic, Miroslav Begovic, 2003. Voltage stability Protection and Control Using a Wide area Network of Phasor Measurements. *IEEE Trans. Power Systems*, 18, 121-127.

[8] Li Fang, Guo Jian, Wu Zhongxi, 2007. Distributed parallel computing algorithms for power system small signal stability based on PC clusters. *Proceedings of the CSEE*, 27, 8-13 (in Chinese).

[9] Chen Shuheng, Li Xingyuan, 2007. An algorithm for identifying low frequency oscillation modes on line based on WAMS. *Relay*, 35, 17-22(in Chinese).

[10] Yang Huimin, Wen Jinyu, 2009. Damping power oscillation on transmission line based on mode-synthesis analysis method. *Automation of Electric Power Systems*, 33, 91-95 (in Chinese).

[11] S. Dutta, B. Wierenga, and A. Dalebout, 1997. Case-based reasoning systems: from automation to decision-aiding and stimulation. *IEEE Trans. Knowledge and Data Engineering*, 9, 911-922.

[12] S. Arya, D.M. Mount, N.S. Netanyahu, D. Silverman, and A.Y. Wu. An Optimal Algorithm for Approximate Nearest Neighbor Searching in Fixed Dimensions. *Journal of the ACM*, 45, 891-923.

[13] P. Zezula, G. Amato, V. Dohnal, and M. Batko, 2006. *Similarity Search–The Metric Space Approach.* Springer, ISBN: 0-387-29146-6.

[14] Wu Cheng, Xu Zheng, Zheng Xiang, 2009. Calculation of Oscillatory Active Power Increment Distribution Based on PSS/E. *Power System Technology,* 2009, 34, 38-42 (in Chinese).

[15] P.Tripathy, S.C. Srivastava, S.N. Singh, 2010. A Modified TLS-ESPRIT Based Method for Low-Frequency Mode Identification in Power Systems Utilizing Synchrophasor Measurements. *IEEE Trans. PowerSystems,* 99, 1-9.

In: Case-Based Reasoning
Editor: Tao Lin

ISBN: 978-1-63482-952-6

Chapter 5

MULTIPLE POWER DISTURBANCES RECOGNITION BASED ON CBR AND COMPLEX WAVELET WITH ADJUSTABLE WINDOW

Tao Lin[1,2,*], Rusi Chen[1,2], Xialing Xu[3] and Guangzheng Yu[1]

[1]School of Electrical Engineering, Wuhan University, Wuhan, China

[2]Hubei Collaborative Innovation Center for High-efficient Utilization of Solar Energy, Wuhan, China

[3]Dispatching Center of Central China Power Grid Corporation, Wuhan, China

ABSTRACT

To realize online recognition of multiple power disturbances in a power grid, a new method is proposed based on Case Based Reasoning (CBR) theory and Continuous Wavelet Transform (CWT) with adjustable complex window function. To obtain pattern features with clear physical meaning, the complex wavelet transform is used to obtain characteristic parameters from electric power waveforms by taking advantage of its good localization properties in both time and frequency domains. Further, CBR is introduced to recognize the types of concurrent multiple power disturbances in electric power waveforms based on the pattern features. Simulation results verify that the CWT can accurately extract characteristic parameters of power disturbances, including harmonics, inter-harmonics, frequency fluctuation and voltage variations. And, the CBR can recognize the types of concurrent multiple power disturbances with high accuracy and efficiency.

Keywords: power quality disturbances, complex wavelet transform, adjustable window, CBR

[*] Corresponding author: School of Electrical Engineering of Wuhan University, Wuhan 430072, China, Email: tlin@whu.edu.cn.

INTRODUCTION

In recent years, extensive application of nonlinear equipments and power electronic devices have caused increasingly serious power quality problems in power grids. Various kinds of power disturbances have exerted serious influences on industrial production. Different power disturbances have different categories and degrees of influence on different users. For example, frequency fluctuation changes the speed of motors in electrical equipment and affects machining precision. On the other hand, voltage change causes maloperations of equipments and further results in the serious loss of electronic precision processing [1]. Thus, it is significant to detect various power disturbance problems and correctly identify the types of power disturbances.

As the basis, extracting the characteristic parameters of each component in a power disturbance waveform is the first step. According to the characteristic parameters and the standard of the classification of power system electromagnetic phenomenon proposed by the IEEE [2], the instantaneous amplitude, instantaneous frequency and duration of each component are usually chosen as the characteristic parameters to conduct power disturbance recognition. The performance of some methods, which are commonly used in power quality detection, are not desirable. For example, harmonic and inter-harmonic components may cause the output increase of RMS algorithm even though there is no amplitude fluctuations of the fundamental component, which leads to the misdiagnosis of voltage swell. The D-q transform method is also affected by harmonic and inter-harmonic components. FS (Fourier Series) method is affected by a non-synchronous sampling, frequency deviation, and harmonic and inter-harmonic are misidentified as voltage fluctuation and flicker due to output oscillations. Hilbert Huang Transform (HHT) suffers from the incompleteness of empirical mode decomposition and is not desirable for the quantitative analysis of signal magnitude-frequency characteristics [3].

In the field of power disturbance recognition, conventional methods such as artificial Neural Network (ANN) and Support Vector Machine (SVM) have various shortages as well. Due to the difficulty in building up ideal training samples and slow training speed for the ANN, efficiency and accuracy of the method proposed in [4] is not desirable. For the method proposed in [5], construction of the SVM based classifier is time consuming. Neither method is able to realize recognition of concurrent, multiple types of power disturbance.

To overcome the above mentioned shortcomings, a new method is proposed based on Case Based Reasoning (CBR) theory and Continuous Wavelet Transform (CWT) with adjustable complex window. The CWT has good dynamic characteristics for measurement of the fundamental component and higher frequency resolution of harmonic components at the same time. The characteristic parameters obtained via CWT are of clear physical meaning. Further, CBR, which has the advantages of easy knowledge acquisition, high reasoning efficiency and strong robustness, is introduced to recognize the types of concurrent, multiple types of power disturbances effectively and efficiently.

The CWT with Adjustable Window Function

Assuming $f(t)$ is a square integral function, $\varphi(t)$ is a basic wavelet function, then the continuous wavelet transform (CWT) of $f(t)$ can be expressed as [8]:

$$C_{cwt}(a,b)=\frac{1}{\sqrt{|a|}}\int_{-\infty}^{+\infty} f(t)\varphi^{*}(\frac{t-b}{a})dt \tag{1}$$

where, $C_{cwt}(a,b)$ is the continuous wavelet transform of $f(t)$; a, b are scale factor and shift factor of the basic wavelet function, respectively, while $\varphi^{*}(t)$ is the complex conjugate function of $\varphi(t)$.

As a Complex Wavelet Function, Morlet wavelet is a single frequency sine function with Gaussian envelope. By taking advantage of concentrated frequency property, narrow pass band and slight band aliasing effect [9], Morlet wavelet function can be used to accurately obtain the amplitude and phase of signal components in power disturbance waveforms. Thus, Morlet wavelet function is selected as the basic wavelet function in this chapter. The time domain expression of Morlet wavelet is:

$$\varphi(t)=\frac{1}{\sqrt{2\pi}}\exp(j\omega_0 t-\frac{t^2}{u}) \tag{2}$$

where u is the bandwidth parameter of the Gaussian window function in the Morlet wavelet. Through dimension scaling transform and displacement translational transform, the following formula can be obtained:

$$\varphi_{a,b}(t)=\frac{1}{\sqrt{a}}\varphi(\frac{t-b}{a}) \tag{3}$$

where $\varphi_{a,b}(t)$ is a wavelet function derived from Morlet wavelet. Provided $w_0=2\pi f_0$ is the central frequency of the Morlet wavelet, the central frequency of wavelet function $\varphi_{a,b}(t)$ is $w=w_0/a=2\pi f_0/a$. $\varphi_{a,b}(t)$ has compact support set in time-frequency plane, its equivalent time-frequency window can be expressed as:

$$(t_0 \pm a\sigma_\varphi)\times(w_0/a \pm \sigma_\Psi/a) \tag{4}$$

where σ_φ and σ_Ψ are the window width of Morlet wavelet in time domain and frequency domain, respectively. Obviously, σ_φ is proportional to u and σ_Ψ is the inverse proportion to u.

Thus, once a is determined according to the central frequency of $\varphi_{a,b}(t)$, the equivalent time-frequency window is not adjustable. On the other hand, the equivalent time-frequency window can be adjusted via u rather than a. To achieve good dynamic performance in the measurement of the fundamental component, u shall be relatively small. To improve frequency resolution in the measurement of harmonic components, u shall be relatively large.

In this chapter, u is 2 and 8 in the measurement of fundamental component and harmonic components, respectively.

OBTAINING PATTERN FEATURES OF POWER QUALITY DISTURBANCES USING CWT WITH ADJUSTABLE WINDOW

CWT with adjustable complex window can be performed to accurately obtain the amplitude A and phase H of signal components in power disturbance waveforms via frequency or scale factor sweeping [10]:

$$\begin{aligned} A &= C\sqrt{a}\left|C_{wt}(a,b)\right| \\ H &= \arctan\frac{\mathrm{Im}(C_{wt}(a,b))}{\mathrm{Re}(C_{wt}(a,b))} \end{aligned} \tag{5}$$

Where C is a constant related to the basic wavelet function and the instantaneous frequency of the signal component, which can be obtained as:

$$f = \frac{1}{2\pi}\frac{dH}{dt} \tag{6}$$

Via frequency sweeping, the amplitudes and instantaneous frequencies of signal components in power disturbance waveforms can be obtained according to the local maxima of their $\left|C_{wt}(a,b)\right|$. Further, CWT is performed via adjusting shift factor b, i.e., via moving time domain window in each sampling interval Δt. The obtained amplitude and frequency of each component in a power disturbance waveform form the following $2m\times n$ matrix X :

$$X = \begin{bmatrix} u_{11} & u_{12} & \cdots & u_{1j} & \cdots & u_{1(n-1)} & u_{1n} \\ f_{11} & f_{12} & \cdots & f_{1j} & \cdots & f_{1(n-1)} & f_{1n} \\ \vdots & \vdots & \ddots & \vdots & \ddots & & \\ u_{m1} & u_{m2} & \cdots & u_{mj} & \cdots & u_{m(n-1)} & u_{mn} \\ f_{m1} & f_{m2} & \cdots & f_{mj} & \cdots & f_{m(n-1)} & f_{mn} \end{bmatrix} \tag{7}$$

Where m is the number of signal components in a power disturbance waveform, n is the number of Δt. u_{ij} and f_{ij} $(i=1,2,\cdots,m; j=1,2,\cdots,n)$ represent the amplitude and frequency of the ith signal component in the jth interval, respectively.

In order to distinguish short-term voltage fluctuation and long-term voltage change, the time accumulation parameter is set for the amplitude and frequency in each column of matrix X. If the change between u_{ij} and $u_{i(j-1)}$ is small, the corresponding time accumulation parameter of jth column is that of $(j-1)th$ column plus Δt. Otherwise, the corresponding time accumulation parameter is Δt. The time accumulation parameters of all columns form a matrix:

$$\tau = \left[t_{i1}, t_{i2}, \cdots, t_{ij}, \cdots, t_{in}\right] \tag{8}$$

Matrix X and τ are used as pattern features for the identification of power disturbances.

RECOGNITION OF MULTIPLE POWER DISTURBANCES BASED ON CBR

CBR is an important reasoning method accompanied with the advance of cognitive psychology in artificial intelligence field. CBR is proposed on the premise that similar situations lead to similar results and can be solved by similar methods [11-13]. Recognition method for multiple power disturbance based on CBR is to identify the types of power disturbances by searching the most similar historical cases. The types of power disturbances can be obtained according to the matched cases and matched results. CBR is composed of the following steps: case representation and organization, searching and matching, reuse and modification, new case addition.

For power disturbance recognition, a case base can be expressed as a series of characteristics, which can be divided into feature attribute and conclusion (types of power disturbances):

$$I_i = < fs_i, c_i > \tag{9}$$

where $fs_i = \{fa_{i1}, fa_{i2}, \cdots, fa_{im}\}$ is the feature attribute of the ith case. According to the standard of classification of power disturbance types proposed by the IEEE, the elements of feature attribute are as follows:

1) Amplitude of the fundamental component.
2) Duration time of the fundamental component.
3) Amplitude of a harmonic component
4) Duration time of a harmonic component.
5) Frequency of the fundamental component.

As the core of CBR, case searching and matching is used to find the most similar case in historical cases base according to the similarity:

(1) For the power quality event $fs_x = \{fa_{x1}, fa_{x2}, \cdots, fa_{xm}\}$ to be recognized, similarity of the kth feature attribute element is measured as:

$$s_{ik} = w_k (fa_{xk} - fa_{ik})^2$$

where w_k is the normalized weight of kth feature attribute element, fa_{xk} and fa_{ik} are the kth feature attribute element of fs_x and fs_i, respectively.

(2) The overall similarity $S(fs_x, fs_i)$ of fs_x and fs_i can be obtained as:

$$S(fs_x, fs_i) = \frac{1}{1 + \sum_{k=1}^{m} w_k (fa_{xk} - fa_{ik})^2} \tag{10}$$

where $\sum_{i=1}^{m} w_k = 1$ must be satisfied.

(3) Repeating steps (1) and (2) for all historical cases in the base to recognize the types of power disturbances in the event according to the most similar cases.

CASE STUDIES

In this chapter, a waveform containing concurrent multiple types of power disturbances, such as voltage sag, frequency fluctuation, harmonic and inter-harmonic, is used to verify the correctness and superiority of the proposed method based on CBR. The waveform is expressed as:

$$u(t) = A_1 \sin(2\pi ft + \frac{\pi}{3}) + A_2 \sin(2\pi ft \times 3 + \frac{\pi}{4}) + A_3 \sin(2\pi ft \times 7.5 + \frac{\pi}{10}) + R_N \tag{11}$$

Where frequency of the fundamental component f is $49.53Hz$. Amplitude of the fundamental component A_1, $3rd$ harmonic component A_2 and $7.5th$ inter-harmonic component A_3 are $200V$, $45V$ and $10V$, respectively. R_N is white Gaussian noise, its mean value and variance are 0 and 0.01, respectively. For simplicity, sampling frequency and the number of samples are $1000Hz$ and 900, respectively. During $[0.3s, 0.6s]$, A_1 decreases to $100V$, while A_2 and A_3 remain the same values.

Table 1. The characteristic parameters of each signal component in the waveform

Time interval	Δt	Δt	Δt	Δt	Δt	Δt
Fundamental Amplitude/V	200	200	100	100	200	200
Fundamental Frequency/ Hz	49.53	49.53	49.529	49.53	49.53	49.53
Harmonic Amplitude/V	44.999	44.999	44.998	44.999	44.999	44.999
Harmonic Frequency/ Hz	148.59	148.59	148.59	148.59	148.59	148.59
Inter-harmonic Amplitude/V	9.9999	9.9999	9.9998	9.9999	9.9999	9.9999
Inter-harmonic Frequency/Hz	371.4749	371.4749	371.4749	371.4749	371.4749	371.4749
τ /s	0.01	0.29	0.01	0.18	0.01	0.18

In the aspect of CWT, w_0 in the Morlet wavelet function is set as 2π, the range of a and Δa are $[0.002, 0.1]$ and 0.0001, respectively. By frequency sweeping, three local maxima values are obtained with corresponding scale factors as $a_1 = 0.02019$, $a_2 = 0.00673$, $a_3 = 0.00269$, which illustrate that there are three signal components in the waveform. Thus, the instantaneous amplitude and frequency of each component can be obtained in each sampling interval via formula (5)-(6). Based on the measurement results, matrix X in formula (7) and the time accumulation parameter matrix/vector τ in formula (8) (just for the fundamental component) per a time interval of 10 ms are listed in table 1. Obviously, the measurement results are correct and accurate.

According to the X and τ in table 1, the proposed method based on CBR is used to identify the types of power disturbances. The recognition results in each time interval are listed in table 2.

Table 2. Recognition results of the types of power disturbances in each interval

Time /s	0.01	0.29	0.43	0.59	0.73	0.9
Disturbance type	Frequency fluctuation	Frequency fluctuation	Frequency fluctuation	Frequency fluctuation	Frequency fluctuation	Frequency fluctuation
	Harmonic	Harmonic	Harmonic Inter-harmonic	Harmonic Inter-harmonic	Harmonic	Harmonic
	Inter-harmonic	Inter-harmonic	Voltage sag	Voltage sag	Inter-harmonic	Inter-harmonic

Table 3. Comparison of the recognition results obtained by the three methods

	Recognition results of power disturbances		
	0-0.29s	**0.3-0.6s**	**0.61-0.9s**
The proposed method	Frequency fluctuation	Frequency fluctuation	Frequency fluctuation
	Harmonic	Harmonic Inter-harmonic	Harmonic
	Inter-harmonic	Voltage sag	Inter-harmonic
Method based on ANN	Voltage fluctuation	Voltage sag	Voltage fluctuation
Method based on SVM	Harmonic	Voltage fluctuation	Harmonic

By statistical analysis of the recognition results, the types of power disturbances in the waveform in each time interval and the recognition results obtained using artificial neural network (ANN) and SVM are compared and listed in table 3.

According to tables 2 and 3, it is obvious that the proposed method can correctly recognize whole types of power disturbances in the waveform. On the other hand, the recognition results using the methods based on ANN and SVM are not desirable. Further, even in the case of white Gaussian noise contamination, the proposed method can efficiently and accurately identify multiple types of power disturbances as well.

CONCLUSION

In this chapter, the CWT with adjustable complex window function is used to accurately extract characteristic parameters of power disturbances. These characteristic parameters form the pattern features with clear physical meanings. Further, by using CBR, which has the features of easier knowledge acquisition, higher reasoning efficiency and stronger robustness, a new method is proposed to identify online the concurrent multiple types of power disturbances via these pattern features. Case studies verify the effectiveness and superiority of the proposed method in comparison with the ANN and SVM based methods.

REFERENCES

[1] Wen, Jifeng. & Liu, Pei. (2002). A new method for detection of power quality disturbances [J]. *Proceedings of the CSEE*, *12*, 17-20.

[2] Li, Tianyun., Zhao, Yan. & Li, Nan. (2005). Apply empirical mode de-composition based Hilbert transform to power system transient signal analysis [J]. *Automation of Electric Power Systems*, 2005, *4*, 49-52.

[3] Lin, Tao., & Fan, Zhengwei. (2007). Application of wavelet transform and artificial neural network to power disturbance identification [J]. *High Voltage Engineering*, *7*, 151-153.

[4] Zhan, Yong., Cheng, Haozhong., Ding, Yi feng., et al., (2005). S-transform-based classification of power quality disturbance signals by support vector machines [J]. *Proceedings of the CSEE*, *4*, 51-56.

[5] Xue, Hui. & Yang, Rengang. (2003). A novel method for non-integer harmonics measurement using continuous wavelet transform [J]. *Automation of Electric Power Systems*, *5*, 49-53.

[6] Liu, Guiying. & Su, Shiping. (2007). Measurement of harmonic and inter-harmonic using transform of wavelet and fourier [J]. *High Voltage Engineering*, *6*, 184-188.

[7] Zhang, Yuhui., Chen, Xiaodong. & Wang, Hongyi. (2004). Continuous wavelet-based measuring and classification of short duration power quality disturbances [J]. *Electric Power Automation Equipment*, *3*, 17-21.

[8] Xue, Hui. & Yang, Rengang. (2002). Morlet wavelet based detection of non-integer harmonics [J]. *Power System Technology*, 2002, *12*, 41-44.

[9] Guo, Yanhong. & Deng, Guishi. (2004). Review of case-based reasoning [J]. *Computer Engineering and Applications*, 2004, *21*, 1-5.

[10] Pham, L. & Wong, K. R. (1999). Wavelet-transform-based algorithm for harmonic analysis of power system waveforms [J]. *IEE Proc. Generation Transmission and Distribution*, *3*, 249-254.

[11] Dutta, S., Wierenga, B. & Dalebout, A. (1997). Case-based reasoning systems: from automation to decision-aiding and stimulation. IEEE Trans. *Knowledge and Data Engineering*, *9*, 911-922.

[12] Zhao, Yanwei. (2000). Research on Case-based Reasoning Intelligent CAD for Designing Machining Center [J]. *Journal of System Simulation*, *2*, 142-145.

[13] Wang, Jun., Pan, Xing., Li, Jing., et al., (2006). Knowledge management consultation system based on CBR technologies. *Journal of Tsinghua University (Science & Technology)*, *S1*, 990-995.

In: Case-Based Reasoning
Editor: Tao Lin

ISBN: 978-1-63482-952-6

Chapter 6

TRANSIENT STABILITY ASSESSMENT OF POWER SYSTEM BASED ON CBR TECHNIQUE

Fei Tang*

School of Electrical Engineering, Wuhan University,
Wuhan, China

ABSTRACT

Signal proceeding technology and pattern recognition methods have been introduced to the Smart Grid in order to resolve some complex issues. A novel method is proposed based on the discrete hidden Markov model (DHMM), which is a kind of case based reasoning (CBR) technique, so as to solve the problems in power system dynamic response combining with the time-sequence of the transient stability assessment (TSA). Firstly, the sensitive feature subset is selected from the original feature set based on the relative sensitivity principle. Thus, the variation trend comparison of the features under different disturbances is realized and dimensions of the signature sequence are properly compressed. Secondly, these features would be transferred into a set of scalars using Lloyds coding method in order to meet the DHMM training requirement. The basic principle of the Lloyds coding method is employing tentative mode in order to diminish quantization errors and to acquire the optimal subareas parameters. The method is utilized to encode the subset vectors into a series of discrete numbers which are feasible to training the model. Finally, the DHMM is employed to be a classier. If the probability of the calculated stable DHMM is greater than that of the unstable one, the result will be stable. On the contrary, the system will be unstable. Based on CEPRI-36 System, New England test system and a practical reginal power grid, simulation results proved the validity and effectiveness of DHMM pattern recognition. Meanwhile, this new method needed less training samples compared with some of the common methods such as ANN, boosting Bayesian and SVM to reach an equivalent accurate rate.

Keywords: DHMM, CBR, transient stability assessment, Smart Grid, pattern recognition

* Corresponding Author address: School of Electrical Engineering, Wuhan University, Wuhan, 430072, China. Email: tangfei@whu.edu.cn

INTRODUCTION

Smart Grid and its technologies are going to be under the spotlight among researchers, which include electrical technology, telecommunication technology, automatic control technology and so on. One of the key points of the Smart Grid is focusing on the two flows: one is the power flow and the other is the information flow. These two flows should be modulated flexibly in bi-directions under smart and scheduled algorithms to meet the optimal cost.

So the Smart Grid needs to introduce new technologies in automation and information technologies to meet the higher requirements. The transient stability assessment of power system (TSA) is a significant issue in the academic circles. The power system is a complex, dynamic, and non-linear system which is in the disturbance all the time. The operation and analysis of power system are becoming more complicated and it is hard to predict. As the system scale is going larger with some kinds of renewable energies plugging in, especially under the construction of Smart Grid.

With the development of telecommunication technology, automatic control technology and artificial intelligence technology, and machine-learned methods have been introduced to power system gradually. Compared with the traditional TSA methods such as time domain simulation method [1] and energy function method [2], the machine-learned method is regarded as one of the most promising online methods [3] with less training data offline and quick classified output online.

The mainstream artificial intelligence methods in TSA are ANN [4], SVM [5], decision tree (DT) [6], Bayes classifier and their improvement algorithms [7-10]. However the training methods of these algorithms ignored the time sequence of the disturbances in the power system, they needed huge scale of training data to achieve the model parameters. It is not practical for large power grid because of the huge amount of calculation.

In this chapter, a novel classifier is proposed in TSA based on discrete hidden Markov model (DHMM), which is a kind of case based reasoning (CBR) technique. The processes of disturbances in the power system are identified as 5 states: normal operation, disturbances' occurrence, disturbances' development, system's swing and system's resuming. In order to achieve better performance, the key steps are listed as following:

1. Decrease the candidate features set based on the relative sensitivity method.
2. The selected features are encoded based on the Lloyds algorithm to be transferred into the training data.
3. Train the DHMM to obtain the model parameters.
4. The unknown samples are put into the trained model to classify.

This chapter is organized as follows. Firstly, the DHMM description is illustrated. Then, DHMM solution is proposed with the detailed training process. Finally, three simulations are carried out and the conclusion is drawn.

Discrete Hidden Markov Model Descriptions

Definition of a HMM

HMM is a kind of Markov chains, which has two stochastic processes: one is a Markov chain to describe the state transition, the other is to explain the relationship between the state and the observations in a statistical way. The whole HMM might be represented by the observed values directly because the state transfers are hidden. Figure.1 shows the typical diagram of the HMM, and an HMM can be described with parameters as follows [11]:

(1) N is the state number in a Markov chain. If the states are marked as $\theta_1, \theta_2, \cdots, \theta_N$ and the Markov chain at time instant t is indicated as q_t, we see $q_t \in (\theta_1, \theta_2, \cdots, \theta_N)$.

(2) M is the number of the observation symbols for each. If the observed values are marked as $V_1, V_2, \cdots, V_M$ and the observed value at time instant t is indicated as O_t, we get $O_t \in (V_1, V_2, \cdots, V_M)$.

(3) π is the initial state probability $\pi = \{\pi_i\}$ where

$$\pi_i = P(q_1 = \theta_i), 1 \le i \le N. \tag{1}$$

(4) The state transition probability distribution $A = \{a_{ij}\}_{N \times N}$, where

$$a_{ij} = P(q_{t+1} = \theta_j \mid q_t = \theta_i), \ 1 \le i, j \le N \tag{2}$$

(5) The observation value probability distribution $B = \{b_{jk}\}_{N \times M}$, where

$$b_{jk} = P(O_t = V_k \mid q_t = \theta_j), \ 1 \le j \le N, 1 \le k \le M \tag{3}$$

So, an HMM could be described as $\lambda = (N, M, \pi, A, B)$, and could be abbreviated as

$$\lambda = (\pi, A, B) \tag{4}$$

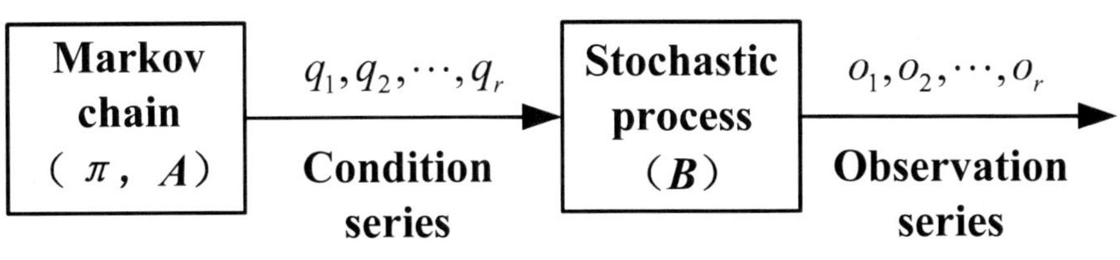

Figure 1. Schematic diagram of the HMM formation.

HMM classifier is first recognized in the speech recognition domain [11], and then researchers employed this algorithm in the hand-shape features recognition [12], in the gait recognition [13], in the text categorization [14], in the transcription factor binding sites [15], etc. In the electrical industry, HMM has been utilized in the deterioration level assessment of power transformers [16], in the nuclear core condition detection [17], in the power disturbance classification [18], in the distribution system reliability assessment [19] and so on. And in this chapter, this classifier would be first applied in the TSA domain.

Transient Stability Assessment Based on DHMM

HMMs can be divided into two kinds of models according to the different observations. One is called continuous HMM (CHMM) with continuous observations and the other is called discrete HMM (DHMM) with discrete observations. Nowadays the coding technologies have been well developed and the continuous signals could be transferred into some kind of digital signals by means of some coding methods. Moreover the quantization error could be under controlled and the digital signals could be easily processed by computers.

The power system transient stability assessment could be regarded as a complicated process by means of pattern recognitions. According to the perspective of time sequence, the process can be identified as 5 states: normal operation, disturbances' occurrence, disturbances' development, system's swing and system's resuming. It is quite difficult to identify every beginning time and ending time of the 5 states so they could be considered "hidden". However the 5 states could be described by some electrical parameters, which means the states could be observed, and such electrical parameters could be regarded as the feature set of the observations. So the DHMM could be introduced to describe such state transitions and the observations are encoded into digital signals. The DHMM had 5 states with the algorithm flow chart as shown in Figure 2.

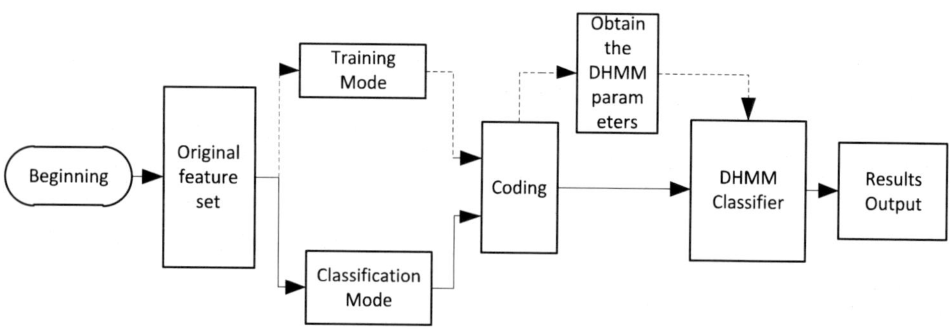

Figure 2. Radical steps for TSA based on DHMM algorithm.

The key step to obtain a DHMM classifier with superior performance is the training method and there are two questions to be resolved: one is how to obtain the electrical feature set which could describe the status at that time; the other is how to code the feature set to make the training easy and efficient.

DHMM Solutions

Key Features Selected Based on Relative Sensitivity Method

57 features are selected to form candidate feature set from the steady-state power flow features. Then the relative sensitivity method is employed to decrease the dimension of the candidate feature set. Function (5) is utilized to calculate the dynamic volume under different disturbances.

$$W_x = \frac{\Delta W}{W_0} \times 100\% = \frac{W_1 - W_0}{W_0} \times 100\% \tag{5}$$

W_0 is the basic value before disturbances , W_1 is the value under disturbances, and W_x is to described as the relative sensitivity. If the $|W_x| \geq 100\%$, this feature would be selected. The feature subset containing 32 features is picked out of the original feature set and the top 32 features are listed as below in the Table 1.

Table 1. Selected feature set in TSA

Item	Features Name
1	the mean of each generator mechanical input power
2	the mean of acceleration active power of all generators
3	the variance of acceleration active power of all generators
4	the mean of relative acceleration active power of all generators
5	the variance of relative acceleration active power of all generators
6	the mean of acceleration of all generators
7	the maximum impact of active power to system
8	the relative maximum impact of active power to system
9	the relative minimum impact of active power to system
10	the relative angle of the generator with a maximum acceleration
11	the maximum acceleration of relative angle of all generators
12	the minimum acceleration of relative angle of all generators
13	the acceleration difference between the leading and the lagging generators
14	the variance of acceleration of all generators
15	the maximum power angle of all generators
16	the minimum power angle of all generators
17	the angle between the minimum and maximum machines among all generators
18	the variance of relative acceleration of all generators
19	the mean of relative acceleration of all generators
20	the maximum rotor kinetic energy of all generators
21	the kinetic energy of the generator with maximum rotor angle
22	the sum of power adjustment of all generators
23	the mean of rotor kinetic energy of all generators
24	the impact of active power to system
25	the sum of the motor speed in COI coordinates

Table 1. (Continued)

Item	Features Name
26	the maximum of the motor speed in COI coordinates
27	the minimum of the motor speed in COI coordinates
28	the angle between the leading and lagging machines among all generators
29	the acceleration between the minimum and maximum machines among all generators
30	the acceleration rate between the minimum and maximum machines among all generators
31	the rotor kinetic energy between the minimum and maximum machines among all generators
32	the rotor kinetic energy rate between the minimum and maximum machines among all generators

Lloyds Coding Method

These 32 features could be regarded as a set of continuous vectors according to the time domain. And these vectors would be transferred into a set of scalars using Lloyds coding method so as to meet the DHMM training requirement. The Lloyds coding method is a mature source coding algorithm in the telecommunication domain [20].

The basic principle of Lloyds coding method is employing tentative mode in order to diminish quantization errors and to acquire the optimal subareas parameters. First, N-1 subareas are divided according to the amplitude and they are arranged in ascending order. Second the values of the 32 features would be mapped into the respective partitions properly. Finally the amplitude x from the vectors set would be coding as a scalar one and the discrete value $index(x)$ could be defined as below:

$$index(x)=\begin{cases}1 & x\le patition(1)\\ i & patition(i)<x\le patition(i+1)\\ N & patition(N)<x\end{cases} \tag{6}$$

where $patition(i)$ is the amplitude of the i subareas, and i represented natural number.

Improvement of the Training Method Based on Multi-observations

Researches [21] deduced the training functions as shown below:

$$\bar{a}_{ij} = \frac{\sum_{t=1}^{T-1} \alpha_t(i)\ a_{ij}\ b_j(O_{t+1})\ \beta_{t+1}(j) / P(O\ /\lambda)}{\sum_{t=1}^{T-1} \alpha_t\ (i)\ \beta_t(j) / P(O\ /\lambda)}$$
$$1 \le i \le N,\ 1 \le j \le N \tag{7}$$

$$\bar{b}_{jk} = \frac{\sum_{t=1,\ \text{且}Q=V_k}^{T} \alpha_t(j)\ \beta_t(j) / P(O / \lambda)}{\sum_{t=1}^{T} \alpha_t(j)\ \beta_t\ (j) / P(O / \lambda)}$$
$$1 \le i \le N,\ 1 \le k \le M \tag{8}$$

where $\alpha_t(i)$ is the forward variable, $\beta_t(j)$ is the backward variable.

However the function (7) and (8) are not suitable to train the DHMM directly, they are for the single observation, not for the multi-observation like 32 features. Moreover, the principle of the HMM algorithm is to calculate the maximum likelihood probability and it is easy to overflow after dozens or hundreds of iterative calculation. The improvement of the training method should resolve the two issues: multi-observation training and overflowing.

L observations series are defined as shown in function (9) where $O^{(l)} = [o_1^{(l)}, o_2^{(l)}, \cdots o_{T_l}^{(l)}]$ is the observations series of l after Lloyds coding and T_l is the sampling length in the time domain.

$$O = [O^{(1)}, O^{(2)}, \cdots O^{(L)}] \tag{9}$$

$$\bar{a}_{ij} = \frac{\sum_{l=1}^{L} \sum_{t=1}^{T_l-1} \alpha_t^{*(l)}(i)\ a_{ij}\ b_j(O_{t+1}^{(l)})\ \beta_{t+1}^{*(l)}(j) / P(O^l\ /\lambda)}{\sum_{l=1}^{L} \sum_{t=1}^{T_l-1} \alpha_t^{*(l)}(i)\ \beta_t^{*(l)}(j) / P(O^l\ /\lambda)}$$
$$1 \le i \le N,\ 1 \le j \le N \tag{10}$$

$$\bar{b}_{jk} = \frac{\sum_{l=1}^{L} \sum_{t=1,\ \text{且}Q=V_k}^{T_l} \alpha_t^{*(l)}(j)\ \beta_t^{*(l)}(j) / P(O^l\ /\lambda)}{\sum_{l=1}^{L} \sum_{t=1}^{T_l} \alpha_t^{*(l)}(j)\ \beta_t^{*(l)}(j) / P(O^l\ /\lambda)}$$
$$1 \le i \le N,\ 1 \le k \le M \tag{11}$$

After the improvement the training functions are deduced as shown in (10) and (11), where $\alpha_t^{*(l)}(i)$ is the scaled forward variable $\alpha_t^{(l)}(i)$ and $\beta_t^{*(l)}(j)$ is the scaled backward variable $\beta_t^{(l)}(j)$. $P(O^l / \lambda)$ is the conditional probability under the consideration of multi-observations.

The Initialization of the DHMM

The TSA process of power system could be identified as 5 states: normal operation, disturbances' occurrence, disturbances' development, system's swing and system's resuming, which is mapped into 5 states of the DHMM (N =5). The state transition probability distribution A is a 5×5 matrix and the observation symbol probability distribution O should contain the 32 features. This kind of time sequence HMM is a left-right model as shown below:

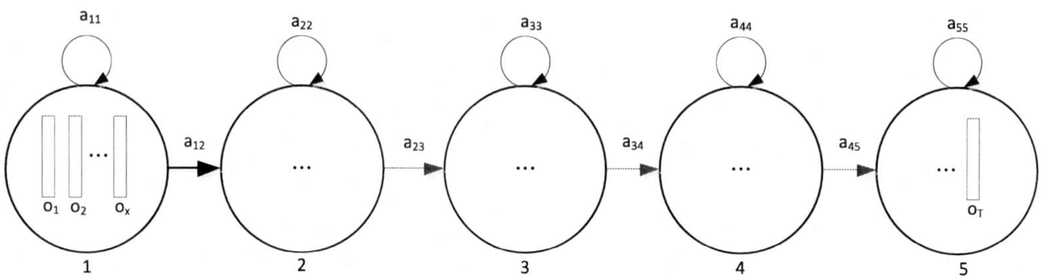

Figure 3. A Special 5-state left-right DHMM.

The initial state distribution π and the state transition probability distribution A could be set as shown in function (12) and (13), and the observation value probability distribution B should be set according normal distribution.

$$\pi = \begin{bmatrix} 1 & 0 & 0 & 0 & 0 \end{bmatrix} \tag{12}$$

$$A = \begin{bmatrix} 0.5 & 0.5 & 0 & 0 & 0 \\ 0 & 0.5 & 0.5 & 0 & 0 \\ 0 & 0 & 0.5 & 0.5 & 0 \\ 0 & 0 & 0 & 0.5 & 0.5 \\ 0 & 0 & 0 & 0 & 1 \end{bmatrix} \tag{13}$$

Process of the Classifier Based on DHMM

The detailed flow of the TSA based on DHMM classifier is shown in Figure 4:

First, the candidate feature set is decreased based on the relative sensitivity method. Second, the selected feature set is normalized and coded based on the Lloyds algorithm in order to obtain the suitable training sequence. Third, the data are divided into two parts: one is the training data including stable and unstable ones, the other is the data to be classified.

Finally the results would come out of the DHMM classifier. Moreover, the flow in the dashed box of Figure.4 showed the detailed training process as follows:

(1) Initial the parameters of the DHMM, such as π, A and B.
(2) Calculate the new parameters of the DHMM based on the BW algorithm [22].

(3) Calculate the optimal conditional probability $P(O|\lambda_j)$ based on Viterbi algorithm [23] with the parameters from (2).
(4) Compared $P(O|\lambda_j)$ with $P(O|\lambda_{j-1})$, if $|P(O|\lambda_j)-P(O|\lambda_{j-1})|\leq\varepsilon$ where ε is set as a threshold ($\varepsilon=10^{-5}$), the iterative would finish because of the convergence.
(5) If the iterative calculation reached the maximum number of the iterations which is set before the beginning without convergence, it would stop and output the parameters.

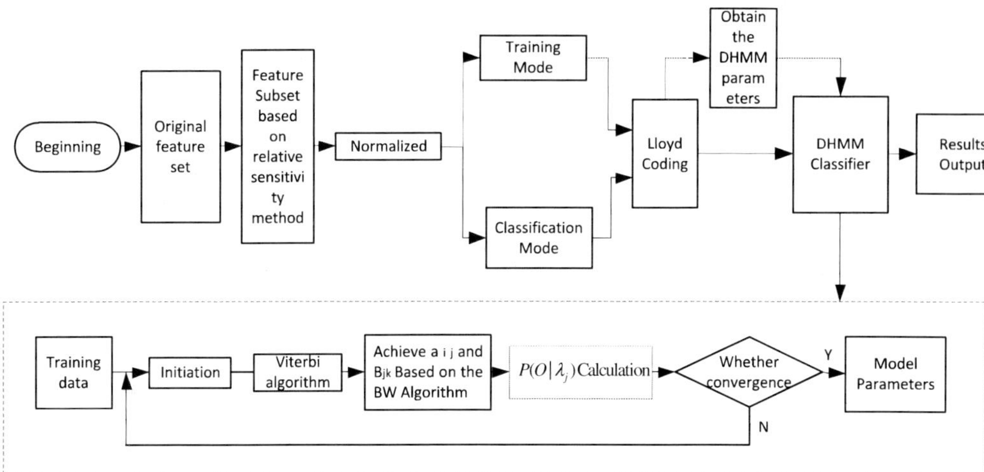

Figure 4. A TSA method based on DHMM algorithm.

CASE STUDIES

The Simulation on CEPRI-36 System

Take the structure of CEPRI-36 system as a simulation example which contained 8 power machines and 36 buses. With the help of Power System Analysis Software Package (PSASP) developed by China Electric Power Research Institute, 300 samples are obtained and 60% of them are taken as the training data while the other 40% are going to be classified by the DHMM. The parameters of the simulation system are set as below:

(1) The power machines are 5-order model taken the excitation system under consideration.
(2) The load is constant impedance load model.
(3) The symmetrical three-phrase short-circuit fault occurred at 0.16s, removed at 0.45s and the total simulation time is 2s.
(4) There are no changes in the topology.
(5) Select one kind of normal operation and take the power flow as a basic one.
(6) Increase the power load 10% as a step from 80% to 130% of the basic load value.
(7) Regulate the generator terminal voltage from 90% to 110%.

All of the generators are under monitoring, if the power angle between any two generators is more than 180 degrees, the system would be unstable. The 300 samples containing the 32 features shown in Table.1 are various from 0s to 4s under different power load.

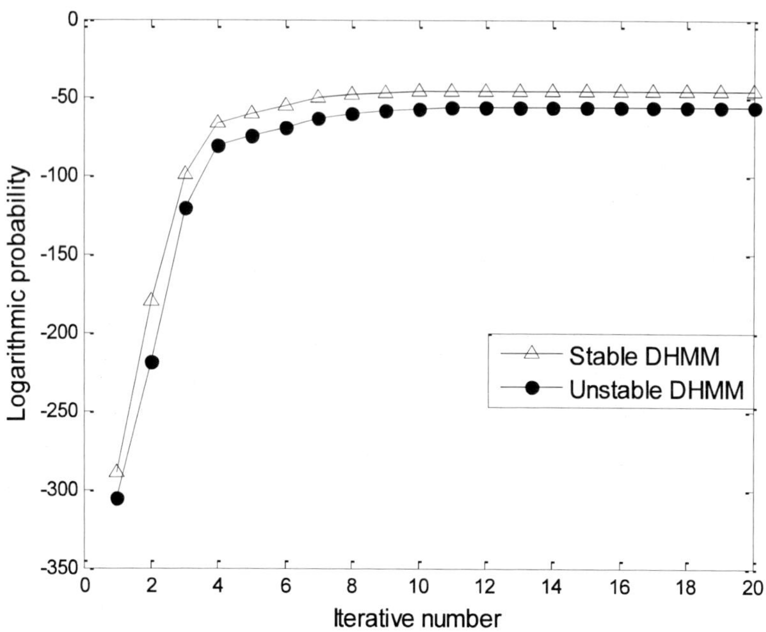

Figure 5. DHMM iterative curves in CEPRI-36 system.

The 32 features are encoded based on the Lloyds method in 2.2, then these coded data are utilized to train the DHMM and the iterative curves had been obtained as shown in Figure 5. The optimal conditional probability could be acquired after more than 15 iterative times, and the parameters of stable and unstable DHMM had been achieved. About 120 samples are pushed into the DHMM classifier to calculate the probability and this chapter also listed four popular artificial intelligence methods under the same simulation system to compare the accuracy rates [6-10] in Table 2.

Table 2. Performance comparison among 5 kinds of algorithms in CEPRI-36 system

Algorithm	*Training samples*	**Classified samples**	*Accuracy rate*
Boosting Bayesian classifier	200	80	97%
Tabu and ANN	400	80	93%
Feature selected and SVM	4678	2589	97.6%
Entropy-based ANN	280	100	95.4%
DHMM	180	120	96.9%

From the Table 2, the DHMM method needed the least training samples to obtain the nearly same accuracy rate among the five classifiers.

The Simulation on Structure of the New England

Take the structure of New England Power Grid as a simulation example which contained 10 power machines and 39 buses [24]. With the help of PSASP, 400 samples are obtained and 50% of them are taken as the training data while the other half are going to be classified by the DHMM. The parameters of the simulation system are set as below:

(1) The power machines are 5-order model taken the excitation system under consideration.
(2) The load is constant impedance load model.
(3) The symmetrical three-phrase short-circuit fault occurred at 0.1s, removed at 0.4s and the total simulation time is 2s.
(4) There are no changes in the topology.
(5) Select one kind of normal operation and take the power flow as a basic one.
(6) Increase the power load 10% as a step from 80% to 130% of the basic load value.
(7) Regulate the generator terminal voltage from 90% to 110%.

All of the generators are under monitoring, if the power angle between any two generators is more than 180 degrees, the system would be unstable. The 400 samples containing the 32 features shown in Table 1 are various from 0s to 4s under different power load.

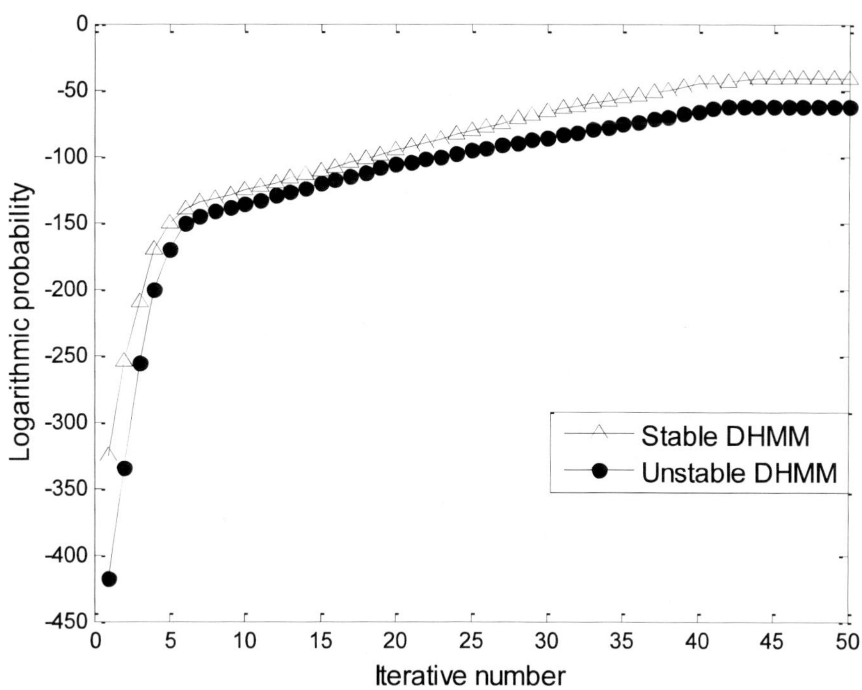

Figure 6. DHMM iterative curves in New England test system.

The 32 features are encoded based on the Lloyds method in 2.2, then these coded data are utilized to train the DHMM and the iterative curves had been obtained as shown in Figure.6. The optimal conditional probability could be acquired after more than 40 iterative times, and the parameters of stable and unstable DHMM had been achieved. About 200 samples are pushed into the DHMM classifier to calculate the probability and this chapter also listed four popular artificial intelligence methods under the same simulation system to compare the accuracy rates in Table. 3.

Table 3. Performance comparison among 5 kinds of algorithms in New England test system

Algorithm	*Training samples*	Classified samples	*Accuracy rate*
Boosting Bayesian classifier	300	100	98%
Tabu and ANN	500	100	95%
Feature selected and SVM	5343	2753	98.5%
Entropy-based ANN	300	100	94.3%
DHMM	200	200	97.5%

From the Table.3, the DHMM method needed the least training samples to obtain the nearly same accuracy rate among the five classifiers.

The Regional Power Grid Simulation System

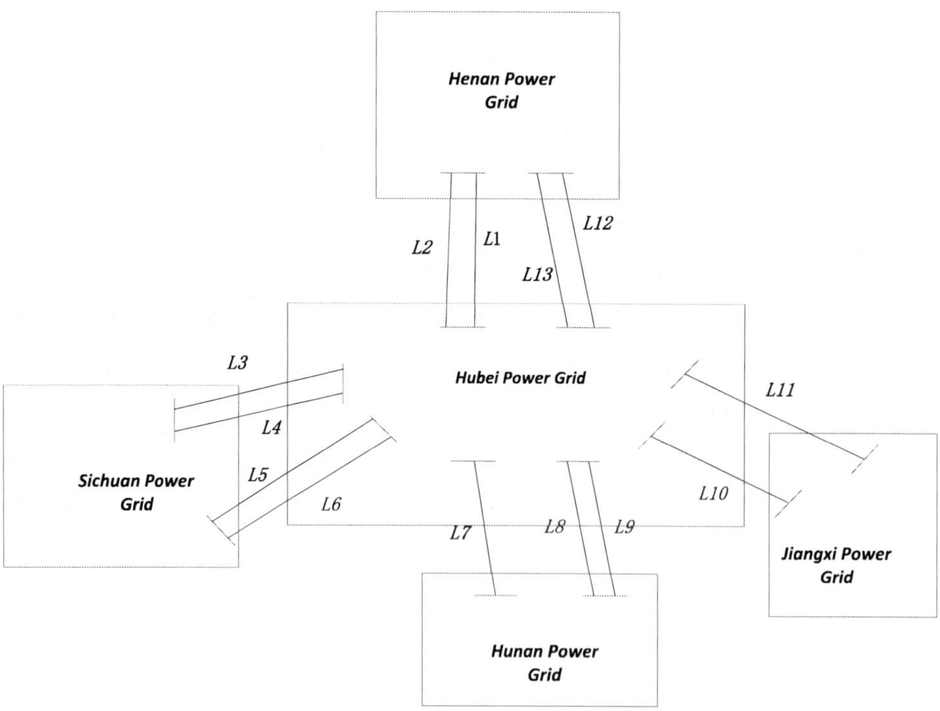

Figure 7. The topology of Central China Power Grid.

The topology of a regional power grid simulation system of Central China is shown in Figure.6 with the power flow of normal operation condition in 2010. The line L1-L2 and L12-L13 are pairs of extra high voltage transmission lines between provinces suffering the heavy load and the system are going to be unstable once there is a short-circuit fault happening. Take the four heavy lines as the researching objects, and the simulation setting is similar to the New England structure. Increase the power load step by step from 80% to 130% uniformly in order to obtain more samples. Similarly, the generator output and terminal voltage are regulated accordingly. 500 samples are obtained from every line and 2000 samples totally. 1000 samples are training data to achieve the DHMM parameters and the iterative curves are shown as in Figure 8.

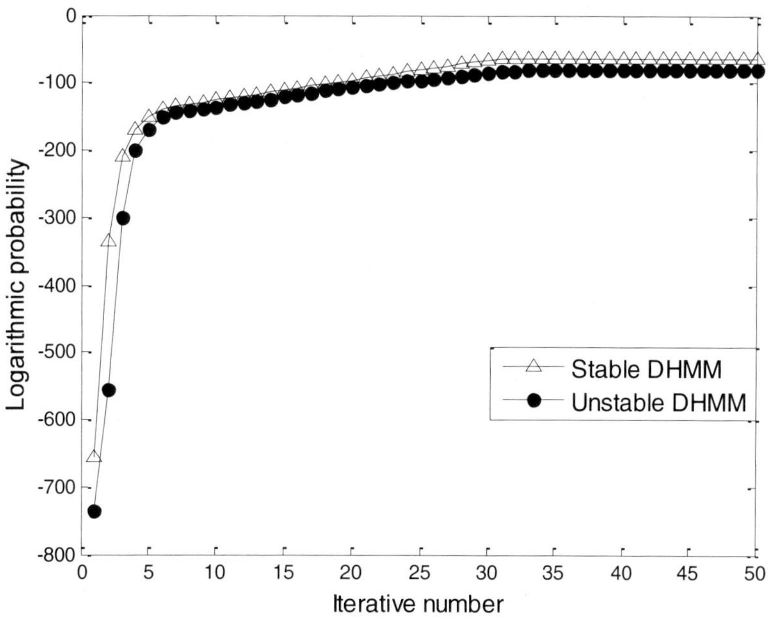

Figure 8. DHMM iterative curves in the real system.

The optimal conditional probability could be acquired after more than 35 iterative times, and the parameters of stable and unstable DHMM had been achieved. The rest 1000 samples are classified by the DHMM. Again, the ANN classifier is employed to compare with the results of the DHMM classifier under the same simulation. Moreover the 32 features are diminished into 10 features based on the principal component analysis (PCA) method in order to reduce the ANN training calculation amount. The results of the two classifiers are listed in Table 4. Obviously, the DHMM method needed less training samples to obtain the nearly same accuracy rate.

Table 4. Performance comparison between DHMM and ANN algorithm in the real system

Algorithm	*Training samples*	*Classified samples*	*Accuracy rate*
PCA and ANN	2896	1104	94.1%
DHMM	2000	2000	96.5%

CONCLUSION

In order to resolve the complicated issues in TSA of the power system, a novel artificial intelligence method is proposed based on DHMM in this chapter. The simulations of CEPRI-36 power system, New England power system and a regional power system are carried out to validate this concept and the conclusions are listed as follows:

(1) It is effective to fit the time sequence of the disturbances in power system, and the five states' exact beginning and ending time in the whole disturbances' occurrence could be ignored.
(2) DHMM method could not only adapt to the small scale power system but also the large power system like Central China Power Grid.
(3) Compared with the traditional and popular method such as ANN, SVM and Bayesian classifiers, this method needs less training samples to reach a relatively high accuracy.

REFERENCES

[1] S. Y. Lee, H.D. Chiang, K. G. Lee, B. Y. Ku. "Parallel power system transient stability analysis on hypercube multiprocessors", *IEEE Transactions on Power Systems,* Vol.6, No.3, pp.1337-1343, Aug 1991.
[2] Michel A, Fouad A, Vittal V. "Power system transient stability using individual machine energy functions". *IEEE Transactions on Circuits and Systems,* Vol. 30, No.5, pp.266-276, 1983.
[3] Savulescu S. C. "Real-time stability in power systems: techniques for early detection of the risk of blackout"[M]. New York, Springer, pp.147-165, 2006.
[4] Amr G. Bahbah, Adly A. Girgis. "New method for generators' angles and angular velocities prediction for transient stability assessment of multimachine power systems using recurrent artificial neural network". *IEEE Transactions on Power Systems*, Vol. 2, No.19, pp.1015-1022, 2004.
[5] L. S. Moulin, A. P. Alves da Silva, M. A. El-Sharkawi, et al. "Support vector machines for transient stability analysis of large-scale power systems". *IEEE Transactions on Power Systems*, Vol. 2, No.19, pp.818-825, 2004.
[6] Wang Kang, Sun Hongbin, Zhang Boming, et al. "Transient stability assessment based on 2D combined attribute decision tree". *Proceedings of the CSEE9*, Vol. 29, No.S1, pp.17-24, 2009.
[7] Lu Jinling, Z hu Yongli, Zhao Hongshan, et al. "Power system transient stability assessment based on boosting bayesian classifier". *Transactions of China Electrotechnical Society*, Vol. 5, No.24, pp.177-182, 2009.
[8] Gu Xueping, Zhang Wenchao. "Feature selection by Tabu search for neural-network based on transient stability classification". *Proceedings of the CSEE,* Vol. 22, No.07, pp.66-70, 2002.

[9] Ye Shengyong, Wang Xiaoru, Liu Zhigang, et al. "Dual-stage feature selection for transient stability assessment based on support vector machine". *Proceedings of the CSEE,* Vol. 30, No.31, pp.28-34.2010.

[10] Li Jun, Liu Yan, Gu Xueping. "An entropy-based discretization method applied in transient stability assessment". Automation of Electric Power Systems, Vol. 8, No.29, pp.26-31, 2005.

[11] Rabiner L R. "A tutorial on hidden Markov models and selected applications in speech recognition" [C]. *Proceedings of IEEE,* Vol. 77, No. 2, pp. 257-285, 1989.

[12] Carlos M. Travieso, Juan Carlos Briceño, Jesús B Alonso. "Transformation of hand-shape features for a biometric identification approach". *Sensors,* Vol. 12, pp.987-1001, 2012.

[13] Changhong Chen, Jimin Liang, Heng Zhao, et al. "Factorial HMM and parallel HMM for gait recognition". *IEEE Transactions on Systems, Man, and Cybernetics,* Vol. 39, No.1, pp.114-123, 2009.

[14] Kairong Li, Guixiang Chen, Jilin Cheng. "Research on Hidden Markov Model-based Text Categorization Process", *JDCTA*, Vol 5, No. 6, pp.244~251, June 2011.

[15] Xiaobao SU, Lifang LIU. "Computational Prediction of Transcription Factor Binding Sites Based on HMM Model and Information Content", *JDCTA*, Vol 5, No. 10, pp.152~159, October 2011.

[16] F. Sotiropoulos, P. Alefragis, E. Housos. "A Hidden Markov Models tool for estimating the deterioration level of a power transformer" [C]. *Proceedings of 12th IEEE International Conference on Emerging Technologies and Factory Automation,* ETFA, Patras, Greece, pp.784-787, 2007.

[17] Bruce Stephen, Graeme M. West, Stuart Galloway, et al. "The use of hidden Markov models for anomaly detection in nuclear core condition monitoring". *IEEE Transactions on Nuclear Science*, Vol. 2, No. 56, pp.453-461, 2009.

[18] Jaehak Chung, Edward J., W. Mack Grady, et al. "Power disturbance classifier using a rule-based method and wavelet packet-based hidden Markov model". *IEEE Transactions on Power Delivery,* Vol.1, No.17, pp.233-241, 2002.

[19] R. E. Brown, S. Gupta, R. D. Christie, et al. "Distribution system reliability assessment using hierarchical Markov modeling". *IEEE Transactions on Power Delivery,* Vol. 4, No.11, pp.1929-1934, 1996.

[20] Winger, L. L. "Linearly constrained generalized LIoyd algorithm for reduced codebook vector quantization". Signal Processing, *IEEE Transactions on,* Vol. 49, No.7, pp. 1501-1509, 2001.

[21] Jinhui Xie. "Hidden Markov Models and Its Application in the Speech Processing"[M]. *Huazhong University of Technology and Science* Press, pp.5-10, 1995.

[22] L. E. Baum, "An inequality and assosiated maximization technique in statistical estimation for probalilistic functions of Markov processes", *Inequalities,* Vol.3, pp.1-8, 1972.

[23] A. J. Viterbi, "Error bounds for convolutional codes and an asymptotically optimal decoding algorithm", *IEEE Trans. Informat. Theory.* Vol. 1T-13, pp.260-269, Apr.1967.

[24] Pai M A. "*Energy function analysis for power system stability*" [M]. Boston: Kluwer Academic Publishers, pp.223-227, 1989.

In: Case-Based Reasoning
Editor: Tao Lin

ISBN: 978-1-63482-952-6

Chapter 7

FAULT DIAGNOSIS EXPERT SYSTEM OF POWER TRANSFORMER BASED ON THE FAULT TREE ANALYSIS

Hui Peng
School of Electrical Engineering, Wuhan University,
Wuhan, China

ABSTRACT

In the process of power system operation, the normal operation of power transformer, which is the one of most crucial equipment in power system, relates to the security and stability of the power system directly. Therefore, the research of diagnosis method for power transformer is an effective and important way to guarantee the stable operation of power system. With the mazy structure and close contact between its components, power transformer has a wide variety of failure modes. Against the above characteristics, this paper will discuss the fault tree diagnostic method for power transformer, and build a power transformer fault diagnosis expert system based on fault tree analysis, so as to diagnosis the power transformer. The method is very logical, intuitive, fully covered and highly reasoning efficiency. it can well meet the requirement of fault diagnosis accuracy, completeness and timeliness in the fault diagnosis of power transformer.

Keywords: power transformer, diagnosis, fault tree analysis, expert system, fuzzy fault tree

INTRODUCTION

As the most critical transmission and transformation equipment in the power system, the operation condition of transformers determines the stability and reliability of the system directly. According to statistics, in the operation, the power transformer is one of the equipment that malfunctions most often. Therefore, we need to timely monitor of the running state of transformers, and diagnose the abnormal state of transformers immediately and accurately.

With the increasing development of electric technology, many fault diagnosis methods, and the intelligent methods of the combined mathematical methods and computer technology have sprung up, such as the three ratio method, non-coding method, BP neural network, grey theory and fuzzy theory. Although all these methods have been used somehow, however, as the electric power industry has entered a period of large grid, large units, high voltage and high automation, voltage level and capacity of power transformers are also constantly increased, and their internal structure has become more complex, more interrelated between components, fault factors are various. Facing these challenges, the shortcomings of each method are also highlighted [1], like the three ratio method, failures that do not exist in coding assembly may occur, and the fuzziness of the boundaries of ratio coding may lead to misjudgment. The BP neural network requires a large number of training samples, which take a longer time of convergent. As a result, these methods can no longer meet the requirements.

There is another diagnostic method that has been extensively applied at present, namely, expert system. Expert system is an artificial intelligence computer program by using the expert's experience, and according to the reasoning method to solve complex problems related to the field. This method has the unique advantage in solving the problem which is hard to establish a mathematical model, and is more dependent on human expert knowledge [2]. But for a large and complex system, the system's high reliability is required, so the traditional expert system based on fuzzy logic rules which is low efficiency, lack of knowledge and accuracy also can't meet the demand of practical diagnosis effectively [3].

Based on the shortcomings and deficiencies of the diagnosis methods above, power transformer fault diagnosis expert system based on the fault tree analysis method is proposed in this paper. Fault Tree Analysis(FTA), is a reliability analysis and fault diagnosis method with strong logic, which is internationally recognized. And it becomes a powerful tool to guide the system optimization design, the weak link analysis and operation maintenance [4], especially for structure, failure mode and the connection between the large complex equipment or system. This article established the power transformer fault model as the form of the fault tree by internal relation between them, and presented logical relationship of all kinds of power transformer fault and its cause visually. The failure modes in the fault tree directly form the rules of diagnosis expert system knowledge base correspondingly, which solves the problem that diagnosis expert system obtains acknowledge uncomfortably, enlarge the coverage of the diagnosis expert system knowledge and also enhance the function of diagnosis expert system. At the same time, the concept of probability importance of minimum cut sets in the fault tree analysis is fused with diagnosis expert system reasoning machine, which greatly improves the search efficiency and accuracy of a diagnosis expert system.

1. Power Transformer Fault Tree Analysis

To analyze power transformer fault tree, the first task is to establish the fault tree of a power transformer, and then realize the corresponding quantitative analysis and qualitative analysis.

1.1. Power Transformer Fault Tree Establishment

1.1.1. FMEA of Power transformer

Power transformer structure is complex, and there are various parts of it. These parts contact closely, which caused the diversity of failure. At the same time, we find that the power transformer fault modes are varied, but the frequency of each failure mode is not the same. Therefore, the "status" of every failure mode in the fault tree is different.

Based on the characteristics above, there are two major problems in the traditional fault tree analysis, which are:

1) In the process of tree establishment, there is no one definite tree-building method for fault tree analysis. Therefore, once the equipment failure relationship is complex and the failure mode is not analyzed systematically, the subjective arbitrariness will be expanded when establishing tree, which makes it impossible to build a relatively complete and reasonable fault tree.
2) In the quantitative analysis of traditional fault tree, generally estimate the probability of bottom events equivalently, which leads to most of the bottom event in the transformer fault tree sharing the same failure probability. Then the failure rate index lost the meaning of distinguishing the bottom events, and the majority of failure mode in the fault tree will have the same "status."

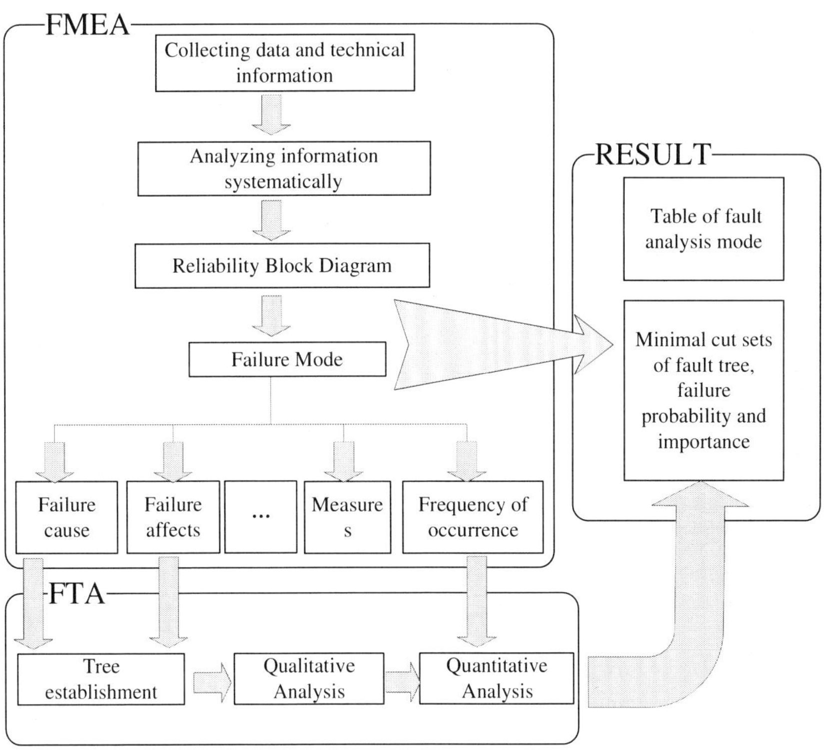

Figure 1. The procedure of FMEA combined with FTA.

each fault subtree separately. Then find out all the minimum cut sets of each fault subtree, which is all the failure modes of each fault subtree. And according to the logical relationship between module fault trees, merge the minimum cut sets of each fault subtree, all minimum cut sets of power transformer fault tree are finally obtained.

Taking advantage of the method above, this paper analyzes the established fault tree of power transformer qualitatively and 117 minimum cut sets are obtained.

Each minimum cut sets corresponds to one independent bottom event. Table 1 shows a part of the minimum cut sets corresponding to the fault tree.

Table 1. Minimum cut sets of fault tree

Top events of fault subtree	Serial number	Minimum cut set
Winding fault	A1	Poor welding
	A2	Short circuit and over stress
	A3	Press cushion block loose
	A4	Short circuit current power
	A5	Wire burr
	A6	Poor welding
	A7	Alignment error
	A8	Partial discharge
	A9	Overload
	A10	Cooling System
	A11	Short circuit current thermal effects
	A12	Natural aging
	A13	Blisters in component or joint possession
	A14	Oil storage tank is sealed and lax
	A15	Silica gel in hygroscopic failure
	A16	The oil pump is sealed and lax

2. Fuzzy Fault Tree Analysis of Power Transformer

2.1. The Fuzziness of Power Transformer Fault Tree Analysis

Traditional power transformer fault analysis is a method of obtaining the probability of top events bottom events by calculating the basic events failure probability. In this process, the failure probability of the bottom events is considered a precise value, however, it is very difficult for the power transformer in the actual situation.

It is mainly based on the following three reasons:

1) The reliability index of equipment, especially as complex as power transformer equipment, usually requires a large amount of data, and these data exist a certain error in most cases.

2) For some high reliability equipment, power transformer for example, it is difficult to get a lot of fault instance data.
3) As for the complex equipment system, it is strongly influenced by human factors or environmental factors. Due to these effects, the specific values of failure probability may vary with circumstances, which cause the inaccuracy of equipment system modeling and the failure of describing the exact probability method.

Therefore, through the analysis above, we will find that when analyzing fault tree of such a large and complex system like power transformers, the fault probability of bottom events is uncertainty, which means there exists fuzziness in power transformer fault tree analysis. The fuzziness is not taken into consideration in the process of traditional power transformer fault tree analysis. The concept of fuzzy number in fuzzy mathematics theory will be introduced in this article to consider the fuzziness of power transformer fault tree analysis.

The so-called fuzzy number is the uncertainty of quantitative processing caused by the concept of fuzziness or various fuzzy factors [11]. The concept of fuzzy number in fuzzy theory is applied to the fault tree analysis, which not only can reduce the difficulty of obtaining the precise value of fault probability of bottom events, but also can build corresponding membership function of fuzzy number considering the experience and the judgment in actual situation of the engineering. This method can describe the precise value of bottom events more accurately. At the same time, the description is also allowed a certain error to make the quantitative analysis becomes more flexible and adaptable.

Fuzzy fault tree analysis is a method of applying the theory of fuzzy mathematics to the fault tree analysis. Now the concept of fuzzy numbers is applied to power transformer fault tree analysis, and here is its basic idea: Firstly, the accurate failure probability of bottom events in classic quantitative analysis is replaced by fuzzy number (F number). And then according to the expansion principle of the fuzzy number in fuzzy mathematics, the OR gate and AND gate fuzzy operator in the fault tree are obtained. Finally we analyze the incident quantitatively according to the fuzzy number and the logic gate fuzzy operator. In this way, we can obtain the probability value range of the incident, which makes the results more close to the actual engineering situation.

2.2. Quantitative Analysis of Fuzzy Fault Tree

In this paper, fuzzy fault tree analysis is mainly focused on the quantitative analysis of fault tree, namely using fuzzy number in fuzzy theory to replace the bottom event fault probability in traditional fault tree analysis. This method cannot only solve the difficulty of precise valuation of bottom events' probability in the traditional method, but also can make the analysis more close to the actual situation by combining practical engineering with technical personnel relevant experience. It's more practical significance.

The general thought of fuzzy fault tree quantitative analysis is as followed. Choose a fuzzy number reference function according to the actual situation of the object to express corresponding fuzzy numbers and to calculate them. It is also the most critical step in the quantitative analysis of a fuzzy number fuzzy fault tree.

After selecting the corresponding reference function, each specific fuzzy numbers of the basic events are calculated combining the research object of the actual data.

Finally, according to the logical relationship of the actual fault tree, the fuzzy number of top incidents is solved with the application of OR gate fuzzy operator and AND gate fuzzy operator, namely, the top event probability is represented fuzzily and the fuzzy probability importance of bottom events is calculated.

2.2.1. The Choice of Reference Function

Common reference function, as aforesaid, mainly includes linear triangular reference function, nonlinear model normal model reference function and pointed reference function. The common reference function [12] represents as follows.

$$\text{triangular}: \begin{cases} L(x)=\max\left\{0,1-\dfrac{m-x}{\alpha}\right\} & x<m, \alpha>0 \\ R(x)=\max\left\{0,1-\dfrac{m-x}{\beta}\right\} & x<m, \beta>0 \end{cases} \tag{1-1}$$

$$\text{normal}: \begin{cases} L(x)=e^{-\left(\frac{m-x}{\alpha}\right)^2} & x<m, \alpha>0 \\ R(x)=e^{-\left(\frac{m-x}{\beta}\right)^2} & x<m, \beta>0 \end{cases} \tag{1-2}$$

$$\text{pointed}: \begin{cases} L(x)=\dfrac{1}{1+\dfrac{m-x}{\alpha}} & x<m, \alpha>0 \\ R(x)=\dfrac{1}{1+\dfrac{x-m}{\beta}} & x<m, \beta>0 \end{cases} \tag{1-3}$$

Each reference function curve is showed in figure 4, 5 and 6.

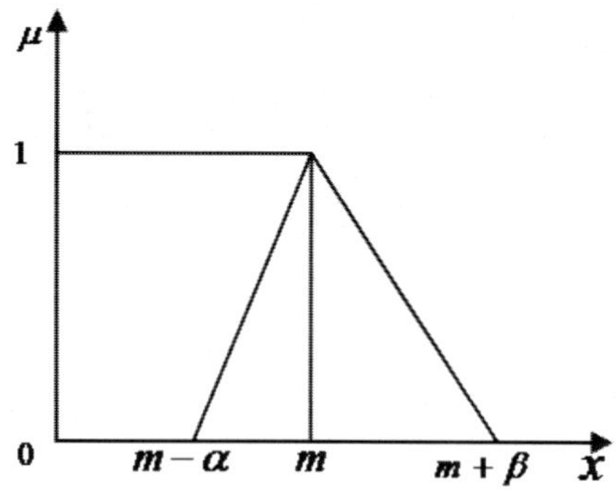

Figure 4. The curve of triangular reference function.

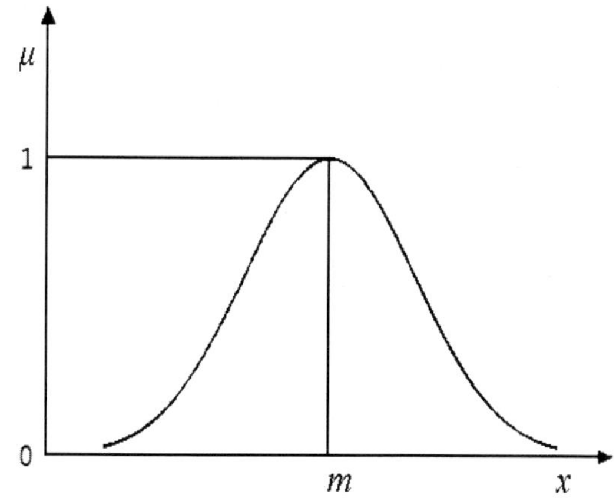

Figure 5. The curve of normal model reference function.

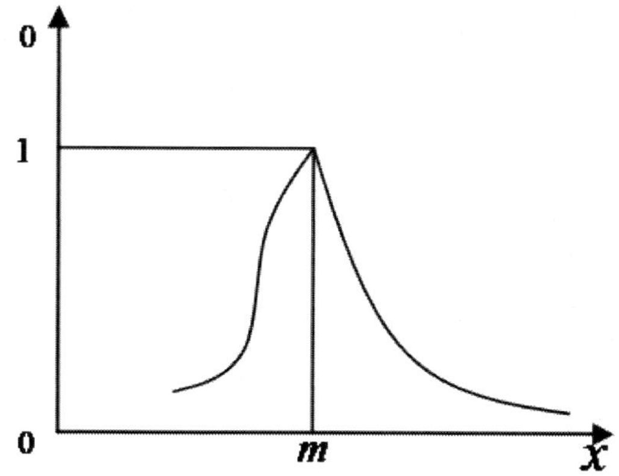

Figure 6. The curve of pointed reference function.

From figure 4, the curve of triangular reference function is linear between "completely belong to" and "completely not belong to". In the section of $x \leq m-\alpha$ or $x \geq m+\beta$, $\mu = 0$, which shows the values in these areas completely do not belong to the fuzzy number. From figure 5 and 6, we find that the both ends of normal type and pointed reference function curves extend to infinity and there are no intersection point between them and horizontal axis, which illustrates that there are subordinate relations between numbers in all the value on the shaft and the fuzzy number to a certain extent.

The choice of reference function directly determines whether the fuzzy fault tree quantitative analysis can proceed smoothly. Therefore, it must be chosen according to some principles:

(1) The fuzzy number reference function of bottom events requires to be more in line with the actual level of fuzzy probability of bottom events. The function would be more close to the truth. This principle is also the most basic principles for selecting reference function.

(2) The calculation error should satisfy the diagnostic requirements. In fuzzy fault tree quantitative analysis, especially the relatively complex ones, the calculation of the fuzzy probability of bottom events and top events requires multiple logical calculations. In the calculation process, calculation error will be accumulated to the next one. The cumulative error will be continuously expanded, and eventually lead to the results becomes meaningless.
(3) In the arithmetic including addition, subtraction, multiplication, and division, the calculation process should be as simple as possible and the calculation should be relatively easy.
(4) In the process of mathematical operation, the structure of the bottom events fuzzy probability should not become more complex. It is best to keep its original mathematical model, which makes the structure of the fuzzy numbers and calculation consequent.

In the actual project, the application of linear triangular fuzzy number and the normal fuzzy number is more comprehensive. Based on the reference function selection principle above, the advantage of the triangular fuzzy number is of smaller computing difficulty, but in comparison, the normal fuzzy number is more reasonable. Article [13] calculates the fuzzy degree of triangular fuzzy number and fuzzy number fuzzy of normal type through the uncertainty analysis, and proves that the triangular fuzzy number is greater than that of normal fuzzy numbers when representing fuzzily the same probability event. Greater fuzzy degree means the bigger uncertainty, so the normal type fuzzy number is better than a triangular fuzzy number. At the same time, the mathematical model keeps constant when performing four arithmetic operations and its calculation error is accord with the requirement of diagnosis. So, the normal fuzzy number is chosen in this article, namely the normal model reference function, to represent bottom events probability fuzzily.

There is certainly subordinate relation between the values of the axis of normal reference function and fuzzy number, and both ends of the curve extend to infinity as shown in figure 5. However, for the sake of convenient calculation, normal fuzzy numbers usually are took cut set processing, and a calculation range will be obtained.

Supposing λ as cut set, we can get the normal type of fuzzy number reference function took cut set:

$$\mu_m(x)=\begin{cases} L(x)=\exp[-((m-x)/\alpha)^2] & ,x\in\left[m-\alpha\sqrt{-\ln\lambda},m\right] \\ R(x)=\exp[-((m-x)/\beta)^2] & ,x\in(m,m+\beta\sqrt{-\ln\lambda}) \\ 1 & ,x=m \\ 0 & ,x<m-\alpha\sqrt{-\ln\lambda}\cup x>m+\beta\sqrt{-\ln\lambda} \end{cases} \tag{1-4}$$

2.2.2. The Fault Tree Fuzzy Quantitative Analysis of Power Transformer

After the selection of fuzzy number reference function, the research object-- Power transformer fault tree can be fuzzily analyzed quantitatively. Power transformer fault tree in this article will be divided into winding, iron core and clamping pieces, insulation, drive pipe, on-load tap-changer, cooling system and non-electricity protection these 7 subtrees. Therefore, each subtree needs to be analysed quantitatively, and then according to the logic relation between each subtree and power transformer, all subtree analysis is incorporated by the way of logic gate fuzzy operator described earlier. Finally the results of the fuzzy fault tree analysis for power transformer are obtained. As mentioned earlier, the reference functions for the occurrence probability P_i of bottom events are the normal type. For the convenience of calculation, take cut sets λ as example, a calculation interval is gained, whose membership function is shown in formula 3-4, and the membership degree of point x which is differ with the mean ±50% is 0.08, so we can get the result as followed.

$$\exp\left[-\left[\frac{m_i - x}{\alpha_i}\right]^2\right] = \exp\left[-\left[\frac{0.5m_i}{\alpha_i}\right]^2\right] = 0.08$$

Table 2. The probability fuzzy number of the basic events

Sunbtree of fault	Serial number	Basic event	Normal fuzzy numbers of probability
winding	A1	Poor welding	(0.002831111, 0.004320827, 0.005430082)
	A2	Short circuit and over stress	(0.002831111, 0.004320827, 0.005430082)
	A3	Press cushion block loose	(0.011564443, 0.017644369, 0.022180649)
	A4	Short circuit current power	(0.011564443, 0.017644369, 0.022180649)
	A5	Wire burr	(0.004607659, 0.007031764, 0.00883751)
	A6	Poor welding	(0.004607659, 0.007031764, 0.00883751)
	A7	Alignment error	(0.004607659, 0.007031764, 0.00883751)
	A8	Partial discharge	(0.006275481, 0.009576492, 0.012036399)
	A9	Overload	(0.003615453, 0.00551774, 0.006934454)
	A10	Cooling System	(0.003615453, 0.00551774, 0.006934454)
	A11	Short circuit current thermal effects	(0.003615453, 0.00551774, 0.006934454)
	A12	Natural aging	(0.003615453, 0.00551774, 0.006934454)
	A13	Blisters in component or joint possession	(0.004085626, 0.006235197, 0.007836247)
	A14	Oil storage tank is sealed and lax	(0.004085626, 0.006235197, 0.007836247)
	A15	Silica gel in hygroscopic failure	(0.004085626, 0.006235197, 0.007836247)
	A16	The oil pump is sealed and lax	(0.004085626, 0.006235197, 0.007836247)

Through calculation, $\alpha_i = \beta_i = 0.3146m_i$ can be obtained. The left and right distribution of each basic event can be calculated from this relationship. The concrete issues are shown in table 2.

According to the probability fuzzy number of the basic events, which is shown in table 2, and combined with the described logic gate fuzzy operator, i.e., using the fuzzy number interval arithmetic algorithm instead of ordinary algorithm, can compute the probability fuzzy number of each fault sub-tree and top event, which is specified in table 3.

Table 3. The probability normal fuzzy numbers of each fault subtree

Subtrees of fault	Normal fuzzy numbers of probability
winding	(0.0768, 0.1103, 0.1426)
Iron core and clamp	(0.0407, 0.0589, 0.0768)
Oil insulation	(0.0273, 0.0397, 0.0519)
Drive pipe	(0.0449, 0.0649, 0.0846)
On-load tap-changer	(0.0509, 0.0735, 0.0956)
Cooling system	(0.0044, 0.0064, 0.0084)
Non-electric quantity protection	(0.0019, 0.0028, 0.0036)

On the premise of fully considering the fuzziness of power transformer, the probability fuzzy numbers of the bottom events and sub-trees are obtained. From all the fuzzy number of fault sub-trees (table 3), winding sub-tree has the highest, then is the on-load tap-changer and drive pipe, which are consistent with the actual fault condition of different parts of the power transformer. Out of preventing power transformers' fault, we need to pay special attention to design, maintenance and maintenance of winding, on-load tap-changer, drive pipe, Iron core and clamp.

3. Expert System Based on Fault Tree Analysis

3.1. The Basic Principle of Diagnosis Expert System

By using the related knowledge, including the actual diagnosis methods and knowledge of expert experience, combined with computer, diagnosis expert system makes the diagnostic knowledge procedural.

According to the corresponding reasoning mechanism, combining with the human-computer interaction interface, user and the computer can do human-computer interaction, and the user makes responses to the suggestions or questions of the system, then system does some inference according to the operation by users.

After completion of reasoning, diagnosis will be given.

The establishment of the diagnosis expert system usually requires a comprehensive understanding of the connection between structure, function of the internal parts being studied.

Because the failure phenomenon, signs and failure itself are not necessarily having a strict correspondence, fault existing at the same time or intermittent failure will increase the construction of corresponding diagnosis expert system. Therefore, master the knowledge of research objects as comprehensive as possible is the key to make the diagnosis expert system get more and better and a reasonable reasoning mechanism.

3.2. Knowledge Acquisition Based on Minimum Cutsets of Fault Tree

Knowledge acquisition methods usually need to do repeated analysis and integration of a large number of examples, data, technical materials and expert experience, and finally form the required knowledge of the knowledge base. This process is very complex and the workload is heavy, and it's difficult to get the logical relationship between various knowledge subjects overall, therefore, it's difficult to obtain accurate and comprehensive knowledge.

A minimum cut set analysed qualitatively by FTA (fault tree analysis) represents a failure mode. From the perspective of knowledge, a minimal cut set of fault tree is equivalent to a piece of knowledge of the system. From the perspective of knowledge representation and structure, each cut set or minimum cut set of the fault tree has a relatively standard knowledge structure.

Based on aforesaid, this article converted all the minimum cut sets to diagnosis knowledge by fault tree analysis, and then a diagnosis expert knowledge system is established. This method is largely reduced the difficulty of the acquisition of expert system knowledge. It is also known as the way of knowledge acquisition based on minimum cut sets of the fault tree.

3.3. Conflict Resolution Strategy Based on the Probability Importance

Conflict resolution strategy is a key part of the expert system reasoning machine. The way to choose one of the knowledge effectively and accurately when there are multiple to choose from will largely determine the inference efficiency of diagnosis expert system. And efficiency is an important reference index to measure diagnosis expert system.

Conflict resolution strategy generally uses a priority ranking to realize the selection of knowledge [15].

Fault tree analysis method calculates the probability importance of the basic events to distinguish the influence degree of the bottom and the top events, which means the greater the probability importance, the greater the likelihood of the bottom events trigger the top events.

So the top events should consider the events of larger important degree first. Based on the concept above, probability importance can be used to do prioritization of knowledge, and then form conflict resolution strategy of a system.

For example, in the drive pipe subtree of Power transformer fault tree, when reasoning processing to "discharge fault → discharge capacitance screen insulation breakdown", there are two pieces of knowledge to choose from, which are “untight end sealing” and “capacitor core manufacturing defects”. Rank them according to the probability importance at this time, the probability importance of “untight end sealing” is greater than the “capacitance core manufacturing defect”, so the system will be more inclined to choose the former. The way to do priority ranking by probability importance to form conflict resolution strategy is also known as conflict resolution strategy based on probability importance.

This method can effectively solve the problem of conflict resolution and improve the reasoning efficiency of the system to a certain extent.

CONCLUSION

As one of the most important electrical equipment in power system, power transformer's health and safely running is related to the reliability and stability of the power system directly.

Therefore, it's meaningful to analyse the fault diagnosis of transformer timely and accurately, and puts forward corresponding suggestions and prevention measures.

In order to handle the characteristics of complex structure and the variety and complexity of transformer, in this paper, a kind of diagnosis expert system based on fault tree analysis is proposed. This method has strong logic, comprehensive coverage and high reasoning efficiency, is good for the requirements of accuracy, completeness, and timeliness of power transformer fault diagnosis. As a result, the proposed method has strong practical application value.

REFERENCES

[1] Cheng Qiming, Tang Wen. The Comparison Research on Three Kinds of Fault Diagnosis of Power Transformer[J]. *Power System Technology*. 2006, 30(Supplementary Issue): 423-425.

[2] Yang Bing. Expert System of Transformer Fault Diagnosis Based on Knowledge Base[J]. *Proceedings of the CSEE*. 2002,10 (22): 121-124.

[3] Liu Xiaoming. The Key Techniques of Launch Vehicle Fault Diagnosis Based on Fault Tree and Neural Network[D]. University of Electronic Science and Technology of China. 2010

[4] Ni Shaoxu, Zhang Yufang, Yi Hong, Liang Xiaofeng. Intelligent Fault Diagnosis Method Based on Fault Tree[J]. Journal of Shanghai Jiaotong University. 2008, 42(8): 1372-1375.

[5] Liu Na, Gao Wensheng. Building and Analysis of Large Power Transformer Fault Tree[J]. *Electric Power*. 2003,36 (11): 33-36.

[6] M. Medarres, G. Deifuli. A truncation methodology for evaluating large fault trees[J]. *IEEE Trans. Reliability*. Vol. R.33.1984,Oct,pp 325-328.

[7] Chen Sanyun. Condition Based Maintenance of Power Transmission and Transformation

[8] Electric Power Industry Ministry of the People's Republic of China. Electric Power Equipment Preventive Test Procedures DL／T596-1 996. Beijing: China Electric Power Press. 1997.

[9] Power Operations of State Grid. Prevention of 110 kv~500 kv Transformer(reactor) and Mutual Inductor Accident Measures Assembly[M]. Beijing: China Electric Power Press. 2003.

[10] Qiu Shiyi. Electric Power Equipment Reliability Maintenance[M]. Beijing: China Electric Power Press. 2004.

[11] Wei Jiazeng. Application and Research of FTA and Fuzzy Theory in Mechanical Fault Diagonosis. Hefei: *Hefei University of Technology*. 2002

[12] Wang Mingjie, Yang Ling. Fault Tree Analysis of Armored Vehicle Autoloader Based on Fuzzy Theory[J]. Ordnance Industry Automation. 2010, 29(5):81-83.

[13] Zhang Xiaoli, Yang Jianqiang, Chang Chunying. Choosing of Fuzzy Number in Fuzzy Fault Tree Analysis Based on Uncertainty Study. *Journal of Chongqing Institute of Technology*. 2005.19(5):19-22.

[14] Li Guixian, Yu Guangbing, Ma liang. Fault Diagnosis to Hydraulic System of Helicopter Landing-fear Based on Fuzzy Fault Tree. *Machine Tool & Hydraulics*. 2007.35(5):214-216.

[15] Guo Weiwei, Ma Jiezhong, Zhai Zhengjun. Research on Ordering of Cut Sets of Fault Tree Analysis [J]. *Computer Engineering and Design*.

In: Case-Based Reasoning
Editor: Tao Lin

ISBN: 978-1-63482-952-6

Chapter 8

FAULT DIAGNOSIS AND RELIABILITY PREDICTION OF HIGH-SPEED RAILWAY TRACTION POWER SUPPLY SYSTEM BASED ON CBR TECHNIQUE

Zhengyou He* *, *Hongbo Cheng and Qi Wang
School of Electrical Engineering, Southwest Jiaotong University
Sichuan, China

ABSTRACT

As the only power source of high speed trains, the Traction Power Supply System (TPSS) is the core part of the High–Speed Railway (HSR). It is essential to ensure safety and reliability of TPSS in order to guarantee proper operation of HSR. In this chapter, research in both short-time and long-time scales are carried out to achieve this goal.

In the short-time scale, fast and accurate fault diagnosis of traction power supply systems will shorten fault treatment time and reduce fault effect. Considering the uncertainty and fuzziness of protection and switches' action information in fault diagnosis, application of Fuzzy Petri Net (FPN), which is a kind of Case Based Reasoning (CBR) technique, in TPSS of HSR fault diagnosis is proposed. Weibull distribution in random mathematics is proposed to model the reliability of relay and circuit breakers, so their reliability indices can be obtained according to the service time. Associating with the confidence of initial placement of FPN, the reliability of relay and circuit breaker makes the confidence determination more practical and more in line with the actual situation. A section of the traction power supply system is proposed as a case to test the method. It shows that the confidence of the diagnosis result varies with the service time of the device, which conforms to reality.

In long-time scale, since environmental meteorological factors have remarkable influences on the status and failure frequency of TPSS of HSR , state prediction and fault prognosis with meteorological information and reliability statistics help to reduce the failure rate or even prevent failure from happening. Meteorological factors are selected by stepwise regression, and the fault and meteorological factor relation model (FMRM) is established by regression analysis. Then the future possible fault under certain conditions

* Corresponding author: State Traction Laboratory of No.111 of North One Section of Second Ring Road, Jinniu District, Chengdu, Sichuan Province, China, 610031. Email: hezy@home.swjtu.edu.cn.

can be predicted by using this model. Reliability function of the traction power supply system is fitted by history fault data, and the reliability at a future specific time can be predicted using this function. Prediction results can provide reference for the operation and management of the traction power supply system, and can provide the basis for preventative maintenance, which is helpful to reduce or avoid the fault.

Keywords: Traction Power Supply System (TPSS), High–Speed Railway (HSR), Fault Diagnosis, Reliability Prediction, Fuzzy Petri Net (FPN), Case Based Reasoning (CBR) technique, Faults And Meteorological Factors Relation Model (FMRM)

INTRODUCTION

By the end of 2014, the total length of the China High-Speed Railway (HSR) reached 12000 kilometers, which accounts for 50% of the world's HSR operation length, ranking top in the world. The length of operating HSR is predicted to reach 16000 kilometers by 2015 and exceeding 20000 kilometers by 2020. The characteristics of China HSR are short construction cycle and large operation scale. The main task in HSR has shifted from large-scale construction to operation and maintenance. As the only power source of HSR, Traction Power Supply System (TPSS) is the key to ensure the reliability and efficient operation of HSR. The majority of abnormal operations and delay are related to the fault of TPSS. Once a failure occurs, the high-speed train would lose its power source and stop running, which would mess up passenger schedules. Even worse, rail transport order would be disturbed, causing tremendous economic loss and reputation damage.

Due to many characteristics of HSR TPSS, it has a high risk of failure. TPSS of HSR has a high risk of failure because of the following features. First, it has a complex structure and harsh operating condition. Secondly, catenary, which is an important component of TPSS, has extensive geographical position and works in the outdoor environment. Moreover, the catenary also needs to bear the continuous high speed impact of locomotive pantograph. Thirdly, power supply conditions change depending on the location of traction substations while climatic conditions and the geographical environment vary from place to place. At the same time, the equipment inside the substation is also impacted by traction load. In order to ensure safe and reliable operation of TPSS and avoid the occurrence of failure, the following two aspects need to be considered: First, fault diagnosis should be done in short time after failure occurs so that we could identify the fault components quickly and reasonably arrange the maintenance activities. Therefore the impact of failure will be controlled in a minimum range. Second, the reliability of the system should be analyzed and evaluated sequentially, which can forecast the occurrence of failure and send alerts. This chapter focuses on the study of the above two aspects.

In terms of the fast fault diagnosis on a short-time scale, this chapter considers the results of the uncertainties of protection devices and breakers information for the fault diagnosis results. Based on the research of the reliability of protection devices and circuit breaker devices variation with time, their Weibull distribution curves are obtained and used to indicate the source of the uncertainties of circuit breakers. Then the relations of fault diagnosis result and the service time of protection and breaker devices are obtained.

For the reliability prediction on a long-time scale, this chapter is based on the historical fault information and meteorological information, from the perspective of the statistics and probability, to evaluate the TPSS reliability under certain conditions. The influence of external meteorological factors on the TPSS operation is analyzed. On the basis of statistical data and regression analysis method, the fault and meteorological factors relation model (FMRM) is established. At the same time, due to the characteristics of the chain structure of TPSS, the Weibull distribution is used to establish the reliability function of TPSS while the statistical data is used to estimate parameters of Weibull distribution. Finally, according to the reliability function, the goal of estimating the reliability of TPSS is achieved.

METHODS

1. Fault Diagnosis

There are many researches on the power system fault diagnosis. These methods are divided into three classes, they are methods based on optimization, methods based on the analysis and calculation and methods based on reasoning. Making full use of various state information from the power grid operation period, these methods use a variety of new intelligent information processing methods to make detection and identification on the grid fault. However, due to the poor working conditions and severe electromagnetic interference of TPSS, many uncertainty factors will exist in the process of information collection, transmission and processing. Therefore, in the process of fault diagnosis, how to deal with the influence on diagnosis, which comes from the uncertainty of the information source to get a more practical and more guiding significance diagnosis, is the need of actual operation.

The Fuzzy Petri Net (FPN), which is developed from Petri Net and is a kind of case based reasoning (CBR) technique, has received more and more attention in the field of power grid and traction power supply system's fault diagnosis, because this model has a rigorous mathematical definition, a intuitional reasoning, a goodness of fit with the process of power grid fault occurrence and treatment, and has good fault tolerance to the uncertainty of data. Especially, FPN utilizes the accuracy of protected action, which it gets from statistical law to represent the real value of protection system and circuit breaker action, and the final diagnosis give a certainty factor in the form of probability, which makes the diagnosis more convincing and the explanation more practical. However, FPN uses long-term statistical average to dispose of the uncertainty of source information and considers the failure rate as constant, which cannot reflect the reliability level at a certain moment, and inconformity with the actual situation of protection devices and circuit breaker actions in the TPSS.

1.1. Fault diagnosis methods of FPN

FPN is developed from basic Petri Nets. This net endows a real number between 0 and 1 to places and transitions, which are used to represent the confidence of places and the possibility of transitions occurrence. A FPN can be defined as:

$$S_{FPN} = (P, T, D, \mathbf{I}, \mathbf{O}, \alpha, \beta, Th, \mathbf{U}, W) \tag{1}$$

where, $P=(p_1,p_2,\cdots p_n)$ is the finite sets of place nodes;

$T=(t_1,t_2,\cdots t_m)$ is the finite sets of transition nodes;

$D=(d_1,d_2,\cdots d_n)$ is the finite sets of propositions;

I : $P\rightarrow T$ is input matrix, which reflects the mapping from places to transitions. $I=\{\delta_{ij}\}$, δ_{ij} is logical variable, $\delta_{ij}\in\{0,1\}$. When P_i is the input of T_j, $\delta_{ij}=1$; When P_i is not the input of T_j, $\delta_{ij}=0$; And $i=1,2,\cdots n$, $j=1,2,\cdots m$;

O : $T\rightarrow P$ is output matrix, $O=\{\gamma_{ij}\}$, γ_{ij} is logical variable. When P_i is the output of T_j, $\gamma_{ij}=1$; When P_i is not the output of T_j, $\gamma_{ij}=0$;And $i=1,2,\cdots n$, $j=1,2,\cdots m$;

$\alpha:P\rightarrow[0,1]$ is the proposition confidence of places, which reflects the places information's credibility;

$\beta:P\rightarrow D$ is a mapping, which reflects the correspondence between place nodes and propositions. If $\alpha(p_i)=y_i, y_i\in[0,1]$, and $\beta(p_i)=d_i$, the confidence of propositions d_i is y_i;

$Th:Th\rightarrow[0,1]$, and the threshold value λ_i is defined for the transition nodes $t_i(t_i\in T)$, $Th=\{\lambda_1,\lambda_2,\cdots\lambda_m\}$;

U: is Certain factor matrix, $U=diag(\mu_1,\mu_2\cdots\mu_m)$, μ_j is the occurrence probability of transition t_j , $\mu_j\in[0,1]$;

$W=W_I\cup W_O$ is the set of input and output weight in all stream relationship, and $W_I=\{w_1,w_2,\cdots w_n\}$, $W_O=\{\mu_1,\mu_2,\cdots\mu_n\}$.

The fault diagnosis, which utilizes FPN in TPSS mainly describes the protection devices and circuit breakers action information's relationship with the component failure action stream, and deduce the true extent of result proposition according to the beginning proposition's confidence coefficient in given rules.

In FPN, if $\sum_j \alpha(p_{Ij})\cdot w_{Ij}>Th(t), j=1,2\cdots n,$ transition t is able to be triggered, and a new confidence coefficient will generate in latter place named P_{Oj}.

$$CF(t)=\begin{cases}\sum_j a(pI_j)\cdot w_{Ij}, & \sum_j a(pI_j)\cdot w_{Ij}>Th(t)\\ 0, & \sum_j a(pI_j)\cdot w_{Ij}<Th(t)\end{cases} \tag{2}$$

This piecewise function named $CF(t)$ could be approximated by a continuous function named $G(x)$.

$$G(x)=x\cdot f_a(x)\cdot\mu \tag{3}$$

where:

$$x = \sum_j a(pI_j) \cdot w_{Ij} \tag{4}$$

$$f_a(x) = \frac{1}{1 + e^{-b(x - Th(t))}} \tag{5}$$

The value of Parameter b is big enough, which makes the formula

$$f_a(x) \approx \begin{cases} 1, x > Th(t) \\ 0, x < Th(t) \end{cases}$$ established.

When the transition in one place satisfies the trigger conditions, this transition will be triggered, and then the confidence coefficient of the next place will become $G(x)$. When multiple transitions are triggered, the new confidence coefficient will choose the biggest results of multiple results generated by multiple input transitions. Then the confidence coefficient of this component will be got. According to the confidence coefficient, the possibility of failure about this component could be judged out.

The power network topology of TPSS in Zhegan Line, Jiangxi province is shown in Figure 1. This section covers five traction substations, which are named North Pingxiang, Xicun, Jieshui, Huangtugang and Zhangshu traction substation. The five traction substations are shown in a dashed box in Figure 1. They can be separated through four partitions named Nuxi, Binjiang, Shuixi and Zhangjiajie, which are represented by CB54, CB57, CB60 and CB63 in Figure 1. This system includes 52 components and 64 breakers. The names of 52 components are: bus A1-A11, B1-B5, line L1-L22, transformer T1-T10, power station G1-G4; The names of 64 breakers are CB1-CB64.

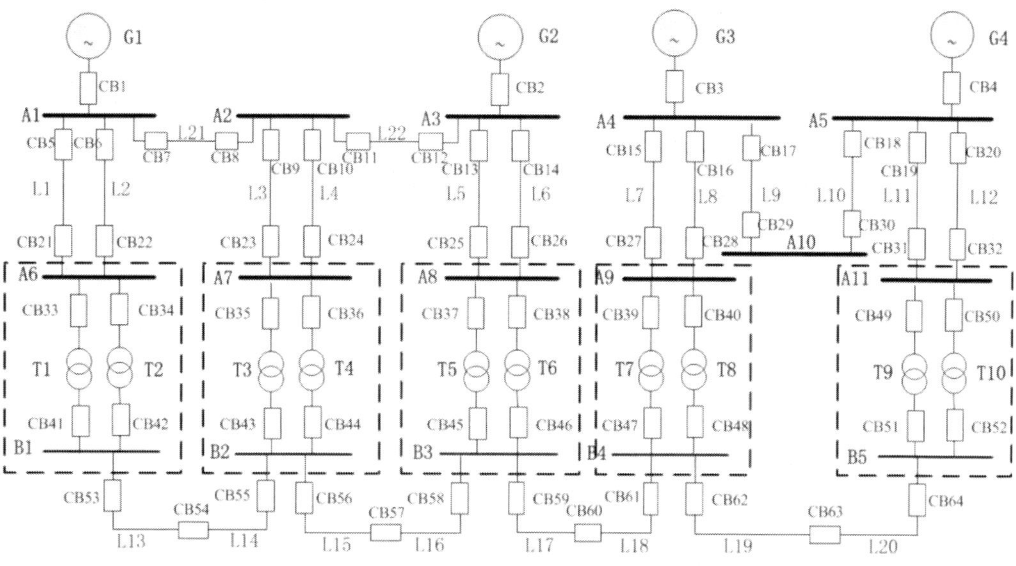

Figure 1. The power network topology of a certain traction power supply system in Zhegan Line.

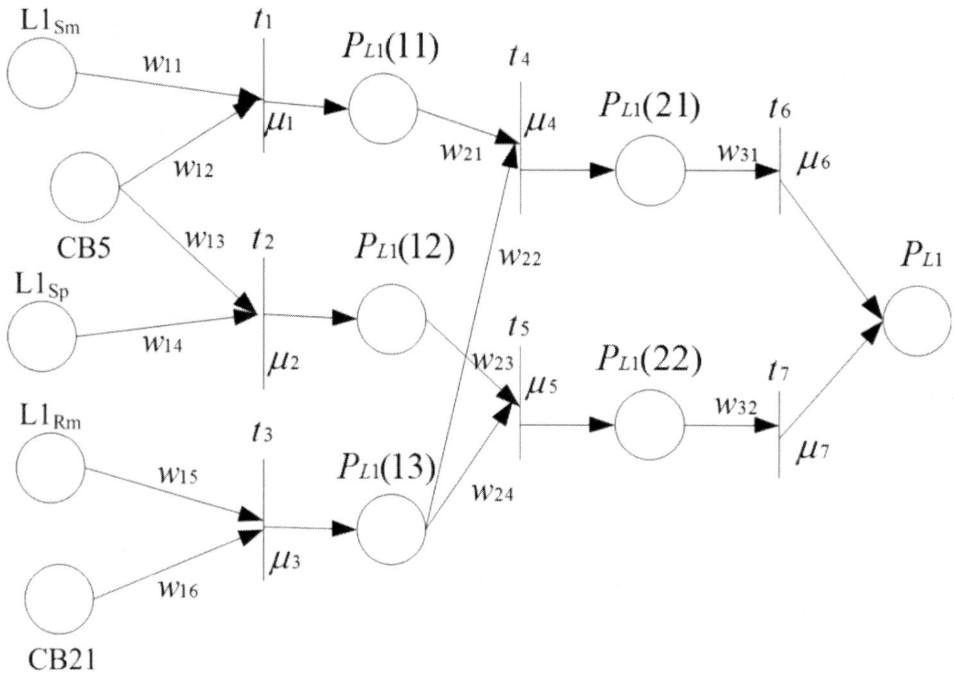

Figure 2. The FPN fault diagnosis model of line L1.

When using FPN to diagnose faults, we could judge the fault component or equipment according to the action information of breakers and protection devices, which are collected by SCADA. For example, if the line L1 malfunctions in Figure 1, we could consider the FPN model, which is composed of main line protection, backup protection and corresponding breaker action, and this model is shown in Figure 2.

In Figure 2, $L1_{Sm}$ and $L1_{Rm}$ represents the upper main protection and bottom main protection of line L1. $L1_{Sp}$ represents the upper backup protection of the line. When line L1 malfunctioned, the main protection $L1_{Sm}$ and $L1_{Rm}$ in the two ends of line will work and disconnect the circuit breaker CB5 and CB21. If the main protection and the corresponding breaker does not work, the line local backup protection $L1_{Sp}$ will work and disconnect the circuit breaker in the two ends of line. w, μ, t in Figure 2 is the same as the definition of FPN in (1).

The common inference method in fault diagnosis is comparing the transition threshold value with the value α in the given initial place. When α is greater than the threshold value, the transition will be triggered and the α value of the next place will be obtained. And by this analogy, the confidence coefficient of the output place will be obtained. The fuzzy inference process shown in Figure 2 is as follows.

1) Firstly, triggering t_1, t_2 and t_3 in T_1.

$$\begin{cases} x_1 = \alpha(L1_{Sm}) \cdot w_{11} + \alpha(CB5) \cdot w_{12} \\ x_2 = \alpha(L1_{Sp}) \cdot w_{14} + \alpha(CB5) \cdot w_{13} \\ x_3 = \alpha(L1_{Rm}) \cdot w_{15} + \alpha(CB21) \cdot w_{16} \end{cases} \tag{6}$$

Using function approximation method to get the α value of the middle library:

$$\begin{cases} \alpha(P_{L1}(11)) = x_1 \cdot \mu_1 / (1 + e^{-b(x_1 - Th(t_1))}) \\ \alpha(P_{L1}(12)) = x_2 \cdot \mu_2 / (1 + e^{-b(x_2 - Th(t_2))}) \\ \alpha(P_{L1}(13)) = x_3 \cdot \mu_3 / (1 + e^{-b(x_3 - Th(t_3))}) \end{cases} \tag{7}$$

2) Secondly, triggering t_4 and t_5 in T_2.

$$\begin{cases} x_4 = \alpha(P_{L1}(11)) \cdot w_{21} + \alpha(P_{L1}(13)) \cdot w_{22} \\ x_5 = \alpha(P_{L1}(12)) \cdot w_{23} + \alpha(P_{L1}(13)) \cdot w_{24} \end{cases} \tag{8}$$

Then,

$$\begin{cases} \alpha(P_{L1}(21)) = x_4 \cdot \mu_4 / (1 + e^{-b(x_4 - Th(t_4))}) \\ \alpha(P_{L1}(22)) = x_5 \cdot \mu_5 / (1 + e^{-b(x_5 - Th(t_5))}) \end{cases} \tag{9}$$

3) Thirdly, triggering t_6 and t_7 in T_3.

$$\begin{cases} x_6 = \alpha(P_{L1}(21)) \cdot \mu_6 / (1 + e^{-b(\alpha(P_{L1}(21)) - Th(t_6))}) \\ x_7 = \alpha(P_{L1}(22)) \cdot \mu_7 / (1 + e^{-b(\alpha(P_{L1}(22)) - Th(t_7))}) \end{cases} \tag{10}$$

4) Finally, the confidence coefficient of output place is obtained.

$$\alpha(P_{L1}) = \max(x_6, x_7) \tag{11}$$

As we can see, the diagnosis process is triggered by the status information of protection devices and circuit breakers. The diagnosis results are related to the confidence coefficient of the status information. Currently, the confidence coefficient about the status information of protection devices and circuit breakers in the initial place could be obtained through the average rate of correct action, which is calculated by statistics.

$$R = \frac{\sum N_r}{N} \tag{12}$$

where N_r represents the number of the device correct action times. N represents the total action number.

The rate of correct action of main protection is 0.9913 according to the literature. The rate of correct action of backup protection is 0.8. The rate of correct action of the circuit breaker is 0.9833. And at the same time, when $b=200$, weight w is 0.5, threshold is 0.3, and rules confidence $\mu=0.95$. The probability of failure in Line L1, which is shown in Figure 2, can be obtained by (6)~(11) as follows:

$$\begin{cases} x_1 = 0.9913\times 0.5 + 0.9833\times 0.5 = 0.9873 \\ x_2 = 0.8000\times 0.5 + 0.9833\times 0.5 = 0.8916 \\ x_3 = 0.9913\times 0.5 + 0.9833\times 0.5 = 0.9873 \end{cases}$$

$$\begin{cases} \alpha(P_{L1}(11)) = 0.9873\times 0.95/(1+e^{-200\times(0.9873-0.3)}) = 0.9379 \\ \alpha(P_{L1}(12)) = 0.8916\times 0.95/(1+e^{-200\times(0.8916-0.3)}) = 0.8417 \\ \alpha(P_{L1}(13)) = 0.9873\times 0.95/(1+e^{-200\times(0.9873-0.3)}) = 0.9379 \end{cases}$$

$$\begin{cases} x_4 = 0.9379 \\ x_5 = 0.8925 \end{cases}$$

$$\begin{cases} \alpha(P_{L1}(21)) = 0.8910 \\ \alpha(P_{L1}(22)) = 0.8479 \end{cases}$$

$$\begin{cases} x_6 = 0.8465 \\ x_7 = 0.8055 \end{cases}$$

And the final diagnosis result is:

$$\alpha(P_{L1}) = \max(0.8465, 0.8055) = 0.8465$$

In other words, the probability of failure in Line L1 is 0.8465.

As we can see from the foregoing description, the features when using FPN for fault diagnosis of TPSS is:

(1) The confidence of diagnosis result is related to the confidence of initial place. In fault diagnosis of power systems, the confidence of the initial place is mostly represented by the correct action rate of protection devices and circuit breakers. However, the correct action rate is closely related to the reliability of the equipment. Therefore, the diagnosis result of FPN should be related to the reliability of protection devices and circuit breakers.

(2) Currently, the confidence of protection devices and circuit breakers in the initial place uses correct action rates, which are calculated based on the statistical information to represent. This rate reflects the long-term average reliability. The average masks the laws of protection devices' and circuit breakers' uncertainty change information, and it cannot reflect the actual situation of protection and circuit breaker real-time, which makes the credibility of fault diagnosis, which can explain the reality greatly reduced.
(3) The weight of protection and circuit breaker information in fault diagnosis lacks of valid basis, and has subjectivity.

As we can see, the reliability of the protection devices and circuit breaker has a greater impact on the results of fault diagnosis. The higher the reliability of equipment is, the higher the confidence that this equipment provides is. At the same time, the weight should be greater in the reasoning process, so as to get a more accurate and realistic diagnosis result.

1.2. Fault model on confidence of place based on Weibull distribution

The reliability of protection devices and circuit breakers are related to their operation time. With the time change, the failure rate and reliability of device is different, therefore the failure rate of the device is a function of time. Weibull distribution is a mathematical model proposed by a Swedish scientist Weibull when he studied the material strength and other issues. Because Weibull distribution has a series of advantages, it is widely used in reliability theory and one of the most widely used models for equipment life and reliability analysis in recent years. With a clear physical background and a lot of practical applications, Weibull distribution has a good compatibility and strong ability to fit the data. Most importantly, it is able to fully describe the failure process and characteristics of different failure periods. Exponential distribution and Normal distribution can be seen as a special case of the Weibull distribution. Moreover, Weibull distribution is suitable for protection devices and circuit breakers reliability modeling.

1.2.1. Weibull distribution of equipment reliability

Failure distribution function of Weibull distribution of two parameters can be determined by the following formula:

$$F(t) = 1 - \exp[-(\frac{t}{\eta})^m] \tag{13}$$

Distribution density function can be calculated as:

$$f(t) = \frac{mt^{m-1}}{\eta^m} \exp[-(\frac{t}{\eta})^m] \tag{14}$$

Reliability function can be calculated as:

$$R(t) = 1 - F(t) = \exp[-(\frac{t}{\eta})^m] \tag{15}$$

Failure rate function can be calculated as:

$$\lambda(t) = \frac{mt^{m-1}}{\eta^m} \tag{16}$$

Where t is the time, m is the shape parameter, which is related to the failure rate trends and η is a scale parameter, which is associated with the lowest failure rate.

Equation (16) implies that if $m<1$, failure rate function $\lambda(t)$ showed a decreasing distribution, failure rate decreases with t and is mainly suitable for modeling initial period failure. If $m=1$, $\lambda(t)$ is a constant, equipment operating in accidental failure period is mainly suitable for modeling random failure; otherwise $m>1$, $\lambda(t)$ showed an increasing distribution, equipment operating in loss failure period and is mainly suitable for modeling aging failure. The different three shape parameters exactly correspond to the three stages of the Figure 3, so the description of the Weibull distribution is generalized.

1.2.2. Parameter Estimation of Weibull distribution

It can be seen that only if shape parameter m and scale parameter η are determined, Weibull distribution is determined. Because the parameters of Weibull distribution are non-linear, they should be converted to linear to apply to the least-squares method. Therefore, take the logarithm on both sides of (13) to get:

$$\ln[-\ln(1-F(t)] = m\ln t - m\ln\eta \tag{17}$$

Let $y = \ln[-\ln(1-F(t))]$, $x = \ln t$, then (17) is simplified to the following equation:

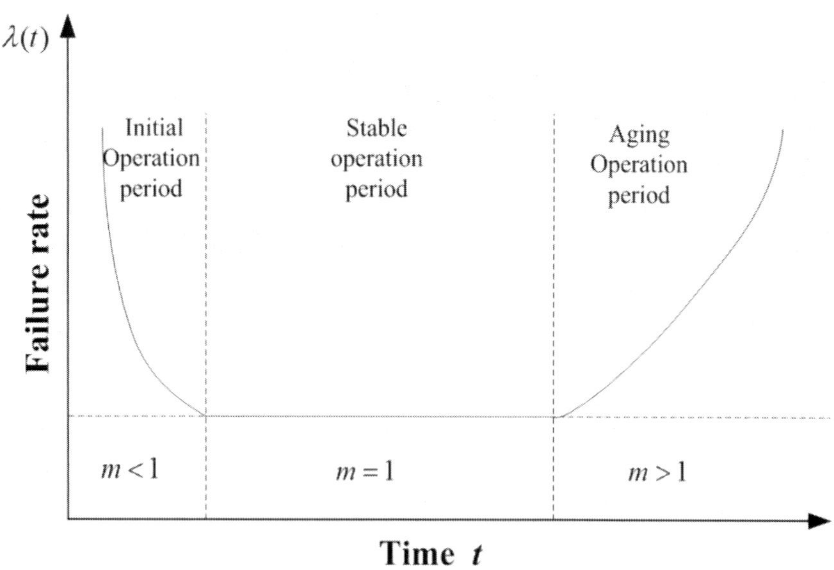

Figure 3. Relations between the bathtub curve of equipment failure rate and the shape parameter.

$$y = mx - m\ln\eta \tag{18}$$

Thus there is a linear relationship. For any samples:

$$\ln[-\ln(1-\hat{F}_i] = m\ln t_{(i)} - m\ln\eta \tag{19}$$

where i is the sequences of failure record and $\hat{F}_i$ is the nonparametric estimate of distribution function $F(t)$. As for the value of $\hat{F}_i$, there are many ways to determine it and the most commonly used is the Bernard median rank estimate :

$$\hat{F}_i = \frac{i-0.3}{n+0.4} \tag{20}$$

where i is the sequences of failure record and n is the sample size of fault data. When the sample size is large enough, the value of $\hat{F}_i$, which is obtained by various methods is very closed. Therefore, the Bernard median rank estimate is applied to estimate the failure distribution function of protection devices and circuit breakers.

After the estimated value of distribution function is obtained, according to the observed sample data, shape parameter m and scale parameter η could be obtained by using the least-square method. Then the expression of equipment reliability function could be obtained.

Least-squares method is based on minimum deviation to estimate parameters m and η. And the deviation is defined as the residual between estimated value and true value. The result is often biased when the independent variable is x or y respectively. So two methods, including minimum vertical residual sum of squares ($(y_i - \hat{y}_i)^2$), and minimum level residual sum of squares ($(x_i - \hat{x}_i)^2$) are used to estimate parameters.

1) Parameter estimation based on the minimum vertical residual sum of squares

Let y be a variable, and the vertical residual sum of squares can be calculated as:

$$R_{yx} = \sum_{i=1}^{r} e_i^2 = \sum_{i=1}^{r} (y_i - \hat{y}_i)^2 \tag{21}$$

Estimated value of shape parameter m and scale parameter η can be calculated as:

$$\begin{cases} m_{yx} = \dfrac{\sum_{i=1}^{r}[(x_i - \overline{x})(y_i - \overline{y})]}{\sum_{i=1}^{r}(x_i - \overline{x})^2} \\ \eta_{yx} = \exp[-(\dfrac{\overline{y}}{m_{yx}} - \overline{x})] \end{cases} \tag{22}$$

where $\hat{y}_i = mx_i - m\ln\eta$, $\overline{x}$ is the average of x and $\overline{y}$ is the average of y.

2) Parameter estimation based on the minimum level residual sum of squares

Let x be a variable, the level residual sum of squares can be calculated as:

$$R_{xy} = \sum_{i=1}^{r} e_i^2 = \sum_{i=1}^{r} (x_i - \hat{x}_i)^2 \tag{23}$$

Estimated value of shape parameter m and scale parameter η can be calculated as:

$$\begin{cases} m_{xy} = \dfrac{\sum_{i=1}^{r} (y_i - \overline{y})^2}{\sum_{i=1}^{r} [(x_i - \overline{x})(y_i - \overline{y})]} \\ \eta_{xy} = \exp[(\overline{x} - \dfrac{\overline{y}}{m_{xy}})] \end{cases} \tag{24}$$

where $\hat{x} = \dfrac{y}{m} + \ln\eta$.

Combining (21) and (23) can get:

$$\frac{R_{yx}}{R_{xy}} = m^2 \tag{25}$$

It can be seen that if $m > 1$, the effect of the obtained parameters is better by the level residual sum of squares minimum model. Otherwise, it is better by the vertical residual sum of squares minimum model.

Table 1. Fault statistics data of a protection device

Basic information	Fault condition			
	Order	**Fault time**	**Operation time/h**	**State order**
Commissioning time 2000-02-01 Deadline 2008-02-01 A total of 21 normal exit record	1	2003-01-01	25560	F1
	2	2003-09-20	31848	F2
	3	2003-03-07	36648	F3
	4	2005-01-01	43104	F4
	5	2005-06-07	46872	F5
	6	2005-10-01	49656	F6
	7	2006-01-01	51864	F7
	8	2006-06-06	55608	F8
	9	2007-01-01	60264	F9

Table 2. Calculation results of Weibull distribution of a protection device

i	t_i	F_i	$x_i = \ln t_i$	$y_i = \ln[-\ln(1-F_i)]$	$x_i - \bar{x}$	$y_i - \bar{y}$	$(x_i-\bar{x})(y_i-\bar{y})$	$(x_i-\bar{x})^2$	$(y_i-\bar{y})^2$
1	25560	0.0230	10.1488	-3.7595	-0.5247	-1.7488	0.9175	0.2753	3.0583
2	31848	0.0559	10.3687	-2.8552	-0.3047	-0.8445	0.2573	0.0928	0.7132
3	36648	0.0888	10.5091	-2.3750	-0.1643	-0.3644	0.0599	0.0270	0.1328
4	43104	0.1217	10.6714	-2.0419	-0.0021	-0.0312	0.0001	0.0000	0.0010
5	46872	0.1546	10.7552	-1.7841	0.0817	0.2266	0.0185	0.0067	0.0513
6	49656	0.1875	10.8129	-1.5720	0.1394	0.4387	0.0612	0.0194	0.1925
7	51864	0.2204	10.8564	-1.3904	0.1829	0.6202	0.1135	0.0335	0.3847
8	55608	0.2533	10.9261	-1.2307	0.2526	0.7799	0.1970	0.0638	0.6083
9	60264	0.2862	11.0124	-1.0873	0.3390	0.9234	0.3130	0.1149	0.8527

Table 3. Weibull distribution estimated parameter contrast of a protection device

Method	Results contrast			
	m	η	residual sum of squares (RSS)	Goodness of fit (R^2)
vertical residual sum of squares	3.0594	83336.843	0.0655	0.9982
level residual sum of squares	3.0932	82740.890	0.0069	0.9999

Table 1 presents the fault recording data of the protection device and Table 2 presents the calculated data.

Table 3 presents the estimated parameters obtained by both level residual and vertical residual, two methods.

It can be concluded from the result that the adoption of the minimum level residual sum of squares method to process parameters estimation will have less residual sum of squares and higher goodness of fitness, which means the fitting performance of the method is more exact than vertical residual sum of squares method. And this is consistent with the conclusion of (25). So the parameters estimation result is obtained by the level residual sum of squares method. The cumulative distribution function and reliability function of the protection device can be calculated as:

$$
\begin{aligned}
F_{\mathrm{re}}(t) &= 1-\exp[-(\frac{t}{82740.890})^{3.0932}] \\
R_{\mathrm{re}}(t) &= \exp[-(\frac{t}{82740.890})^{3.0932}]
\end{aligned}
\tag{26}
$$

Figure 4 presents the reliability function curve of the protection device.

When FPN is used for fault diagnosis, another signal comes from motion information of the circuit breaker. Due to poor working condition of TPSS, motion of circuit breakers is frequent. With the times of motion of circuit breakers increases, wear of breakers contacts and mechanical structure increases, and their reliability is reduced.

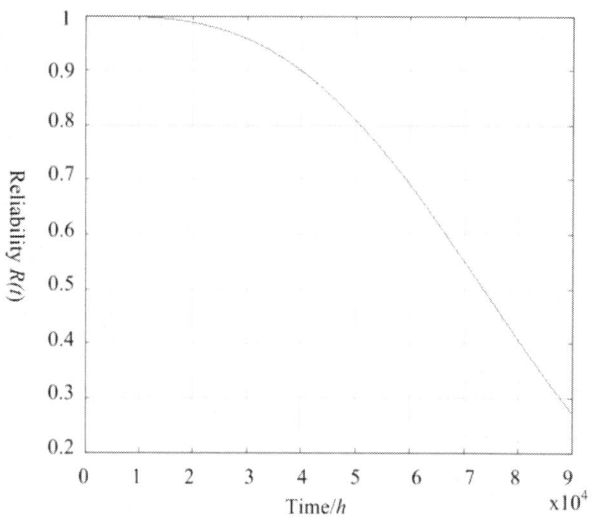

Figure 4. Reliability curve of the protection device.

Table 4. Failure statistics data of a vacuum circuit breaker

Order	Failure times	Cause of the failure	State order
1	10010	Insulated rod deviates from the guide frame	F1
2	18015	Drive shaft failure	F2
3	26000	Over travel beyond the technical scope	F3
4	36483	Permanent magnetic bodies in turn coil burnout	F4
5	42000	Vacuum arcing chamber pressure failure	F5
6	50000	No failure occurs, stop the test	

Here failure statistical data is analyzed to achieve reliability function of the type of vacuum circuit breakers. Table 4 presents the failure statistical data of this type of vacuum circuit breakers.

Differing from protection devices, operating frequency of circuit breakers is regarded as a measure of its life. As for this type of the circuit breaker, statistical time T =50000 and during the experiment the number of failures r =5.

Table 5 presents the calculated data, which is obtained by both level residual and vertical residual methods.

Table 6 presents the estimated parameters obtained by both level residual and vertical residual methods.

Table 5. Calculation results of Weibull distribution of a circuit breaker

i	t_i	F_i	$x_i = \ln t_i$	$y_i = \ln[-\ln(1-F_i)]$	$x_i - \bar{x}$	$y_i - \bar{y}$	$(x_i-\bar{x})(y_i-\bar{y})$	$(x_i-\bar{x})^2$	$(y_i-\bar{y})^2$
1	10010	0.1094	9.2113	-2.1556	-0.8539	-1.3961	1.1921	0.7291	1.9490
2	18015	0.2656	9.7990	-1.1753	-0.2663	-0.4157	0.1107	0.0709	0.1728
3	26000	0.4219	10.1659	-0.6015	0.1006	0.1580	0.0159	0.0101	0.025
4	36483	0.5781	10.5046	-0.1473	0.4394	0.6123	0.2690	0.193	0.3749
5	42000	0.7344	10.6454	0.2819	0.5802	1.0415	0.6043	0.3366	1.0847

Table 6. Weibull distribution estimated parameter contrast of a circuit breaker

Method	Results contrast			
	m	η	residual sum of squares (RSS)	Goodness of fit (R^2)
vertical residual sum of squares	1.6360	37305.7331	0.0203	0.9931
level residual sum of squares	1.6453	37403.3894	0.0075	0.9998

It can be concluded from the result that the adoption of the minimum level residual sum of squares method to process parameters estimation will have less residual sum of squares and higher goodness of fit. Therefore, the parameters estimation result is obtained by the level residual sum of squares method. The cumulative distribution function and reliability function of the circuit breakers can be calculated as:

$$F_{CB}(t) = 1 - \exp[-(\frac{t}{37403.3894})^{1.6453}]$$
$$R_{CB}(t) = \exp[-(\frac{t}{37403.3894})^{1.6453}] \quad (27)$$

Figure 5 presents the reliability function curve of the circuit breakers.

As for above protection devices and circuit breakers, $m > 1$ can be obtained via the parameter estimation, which means both devices are in an aging operation period and their reliability is reduced over time. Therefore, the credibility of signals which are provided by different protection devices and circuit breakers changes over time. When using these signals, the impact of device reliability to source credibility should be considered.

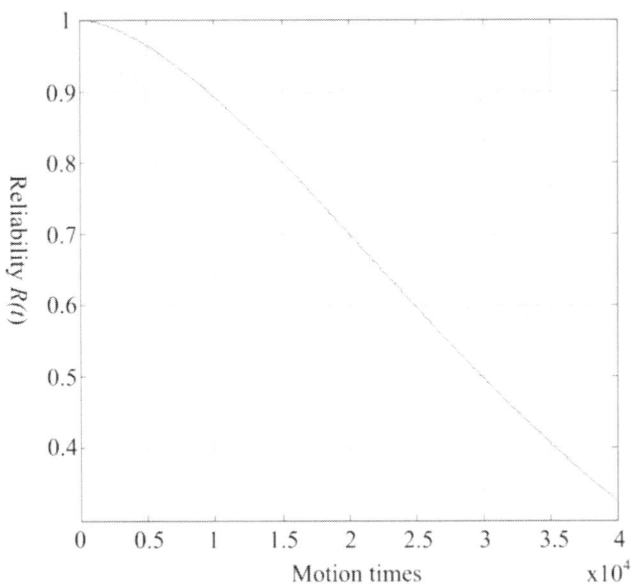

Figure 5. Reliability curve of a vacuum circuit breaker.

1.3. TPSS fault diagnosis considering the signal source device reliability

1.3.1. FPN based fault diagnosis of TPSS

In fault diagnosis using fuzzy Petri, the confidence of initial place α directly affects the credibility of the final diagnosis. α gives the trust of protection devices and circuit breakers information. It is associated with the reliability of the protection devices and circuit breakers. The higher the reliability of the equipment, it provides the higher the credibility of information, and the weight of the information in the process of reasoning is greater. In this section, the FPN information confidence level of the aforementioned method is modified in the process of fault diagnosis. It uses the equipment reliability function to describe the credibility of the equipment to provide information, gets the time relationship between fault diagnosis and equipment service, modifies the information weight in the process of reasoning and decides the information weight according to the degree of reliability. The information provided by the high reliability equipment has great weight in the process of reasoning. Otherwise, the information provided by the low reliability equipment has low weight in the process of reasoning. This makes the reasoning results more credible and can reduce effects of equipment misoperation and refusing action.

In this section, the FPN for fault diagnosis is as follows:

$$S_{FPN} = (P, T, D, \mathbf{I}, \mathbf{O}, \alpha, \beta, Th, \mathbf{U}, W)$$

Its definition is (1), but the difference is that the determination of its confidence of initial place α is not with the method of (12) and its weight w is not artificially random specified.

The initial degree of confidence of protection device place:

$$\alpha(P_{re}) = R_{re}(t) \tag{28}$$

The initial degree of confidence of circuit breaker place:

$$\alpha(P_{CB}) = R_{CB}(t) \tag{29}$$

The weight of each place when reasoning:

$$w(P(i)) = \frac{R_i}{\sum_j R_j} \tag{30}$$

Among them, R_i is the corresponding reliability by the place, $\sum_j R_j$ is the sum of the reliability of all places in the reasoning. For the initial place, R_i can be directly given by equipment reliability function, For the middle place, it can be expressed using equivalent reliability, and the equivalent reliability is decided by the former level place's reliability in reasoning:

$$R_i(P(kl)) - \prod_m R_r(P((k-1)m)) \tag{31}$$

where $R_r(P((k-1)m))$ is the reliability of former level of place, which participates in the place $P(kl)$ reasoning.

In this way, the initial place confidence will be a function of time and change with equipment reliability; The places information will be adjusted dynamically as the change of equipment reliability in the process of reasoning.

Take line L1 fault diagnosis FPN in Figure 2 is an example. The reliability function of the protection device is represented as (26), the reliability of the circuit breaker is represented as (27), Figure 6-Figure 8 show the relationship of all the information weight changes with equipment service time in the fuzzy inference.

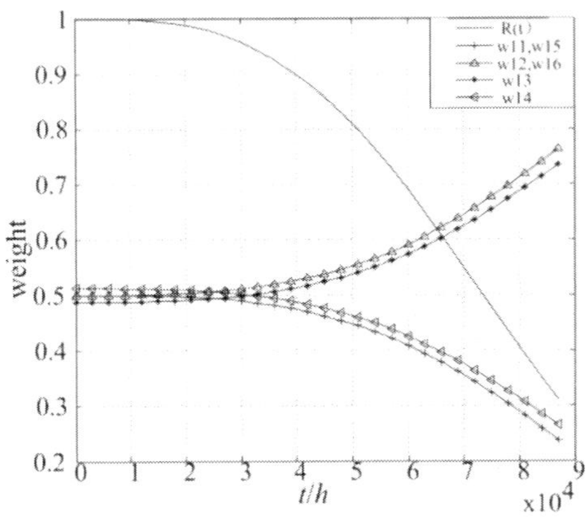

(a) The changing curve of the input place weight w_1 as the protecting operation time (circuit breaker action number t_2=1).

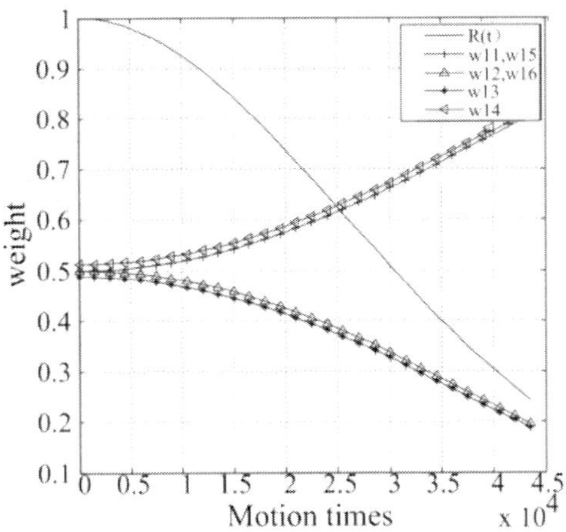

(b) The changing curve of the input place weight w_1 as the circuit breaker motion times (the running time of protection t_1=1).

Figure 6. The input place weight w_1 changes with the time.

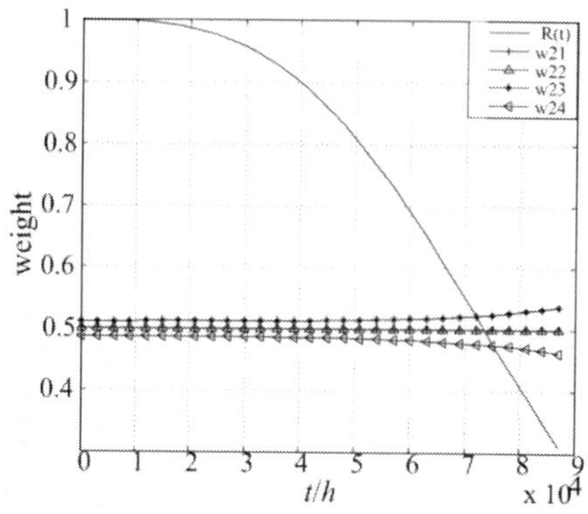

(a) The changing curve of the middle place weight w_2 as the protecting operation time (circuit breaker action number t_2=1).

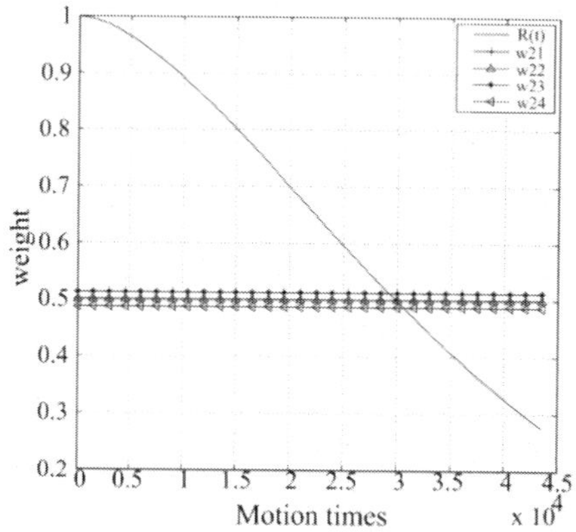

(b) The changing curve of the middle place weight w_2 as the circuit breaker motion times (the running time of protection t_1=1).

Figure 7. The middle place weight w_2 changes with the time.

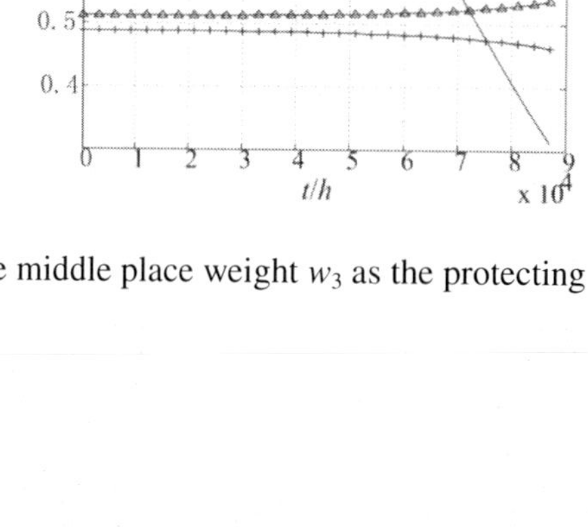

(a) The changing curve of the middle place weight w_3 as the protecting operation time (circuit breaker action number t_2=1).

Figure 8. (Continued).

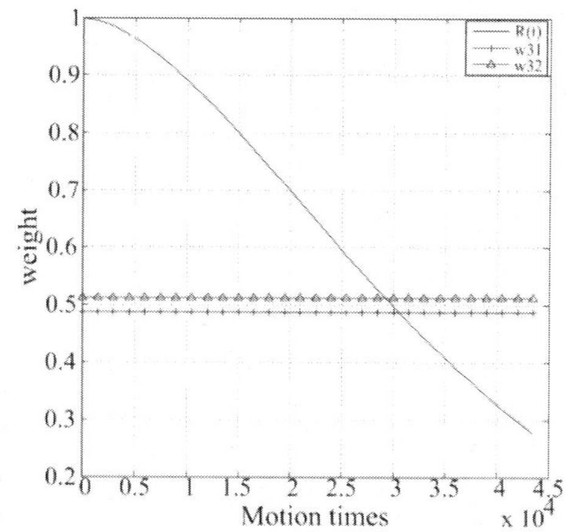

(b) The changing curve of the middle place weight w_3 as the circuit breaker motion time (the running time of protection t_1=1).

Figure 8. The middle place weight w_3 changes with the time.

It can be seen from Figure 6, with the growth of the protection device running time, the reliability of the protection equipment reduces the related weight w_{11}, w_{14}, w_{15} of fault diagnosis and protection information gradually reduces. The weight of the information provided by the protection device gradually reduces, but the weight of circuit breaker information gradually increases. Correspondingly, when circuit breaker action times increase, its reliability gradually reduces and the weight of circuit breaker information in fuzzy reasoning reduces gradually. So this method can dynamically adjust weight of the equipment's information in the fault reasoning according to the degree of the reliability of the equipment and can make the diagnostic results more practical.

Because the middle places $P_{L1}(11), P_{L1}(12), P_{L1}(13)$ are related to the reliability of the protection device, the influence of the protection device reliability changes to them are similar, this makes the changes of subsequent middle places' weight $w_{21}, w_{22}, w_{23}, w_{31}, w_{32}$ less affected by the protection device and circuit breaker reliability, as shown in Figure 7 and Figure 8.

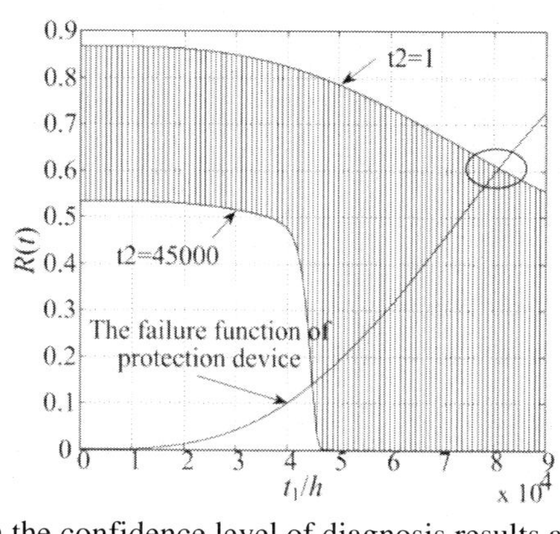

(a) The relationship between the confidence level of diagnosis results and the operation time of protection device.

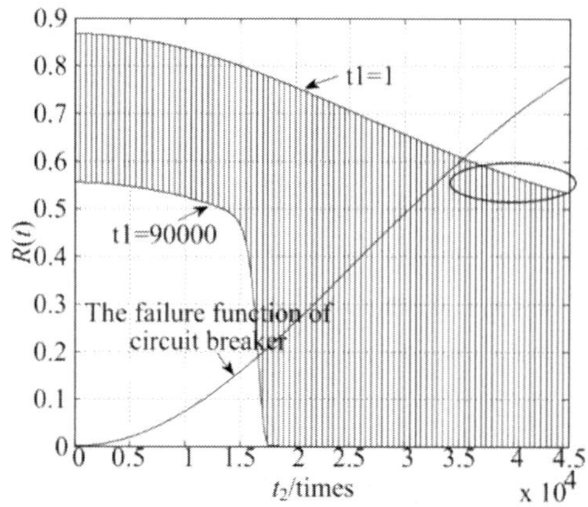

(b)the relationship between the confidence level of diagnosis results and the circuit breaker motion times.

Figure 9. The relationship between the confidence level of diagnosis results and time.

The relationship between the confidence level of fault diagnosis and time is shown in Figure 9; it can be seen that the diagnosis confidence changes with time, when the protection device and the circuit breaker run for a period of time, and after the reliability reduces to a certain degree, the diagnosis confidence will reduce to 0, and reasoning results think the equipment action is a misoperation.

Because the dynamic weight is used, when a certain equipment reliability decreases, the equipment information's weight is reduced. Correspondingly, the weight of other information is improved. So in the later stages of the running time, the case that the credibility of the diagnosis reduce will be improved, as the red oval shows in Figure 9.

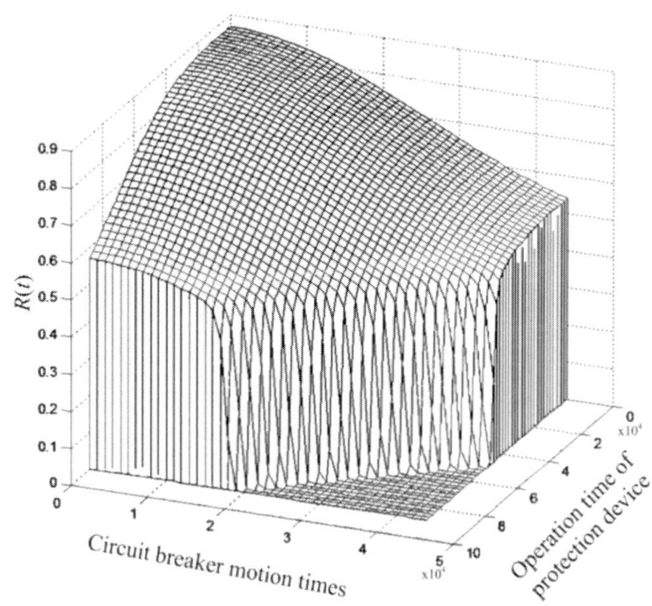

Figure 10. The relationship between the confidence level of diagnosis results and the running time of protection device and the circuit breaker.

Through the above analysis, it can be seen that the diagnosis confidence is affected by the source device - the protection devices and reliability of circuit breakers. Therefore, the diagnosis confidence is related with the running time of protection devices and circuit breakers and in fact, it is the function of both. Figure 10 shows the relationship between the confidence level of fault diagnosis results and the running time of protection devices and the circuit breakers action number of times. It can be seen from Figure 10, when the protection devices and the circuit breakers run for a long time, the reliability of both decreases and the confidence of the fault diagnosis results, which are reached according to the information they provide, also decreases with it.

1.3.2. The Case Study

According to the foregoing, considering the FPN fault diagnosis model for the reliability of the equipment, using the reliability function of protection devices and circuit breakers, which are obtained in 0 as the basis decided by the FPN initial places confidence according to different components in TPSS in Figure 1, the FPN diagnosis model is established. Take typical components for example, bus B1, traction transformer T1, line L13, the established FPN diagnosis model is shown in Figure 11, Figure 12, and Figure 13.

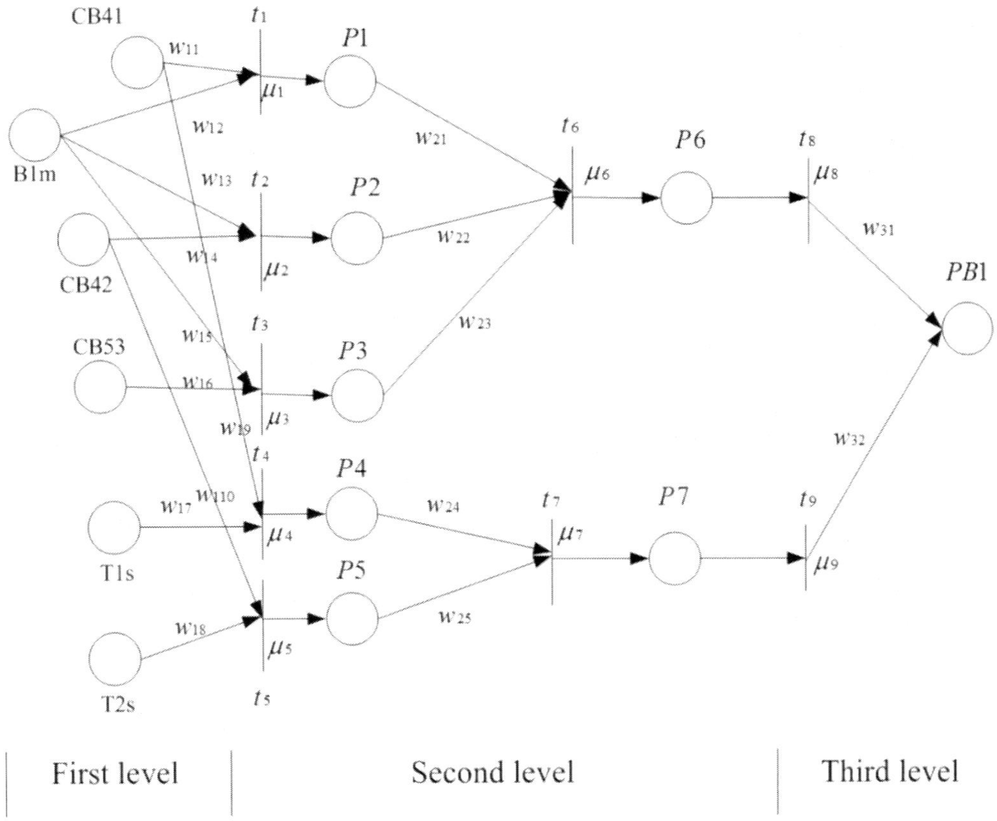

Figure 11. The FPN model of bus B1.

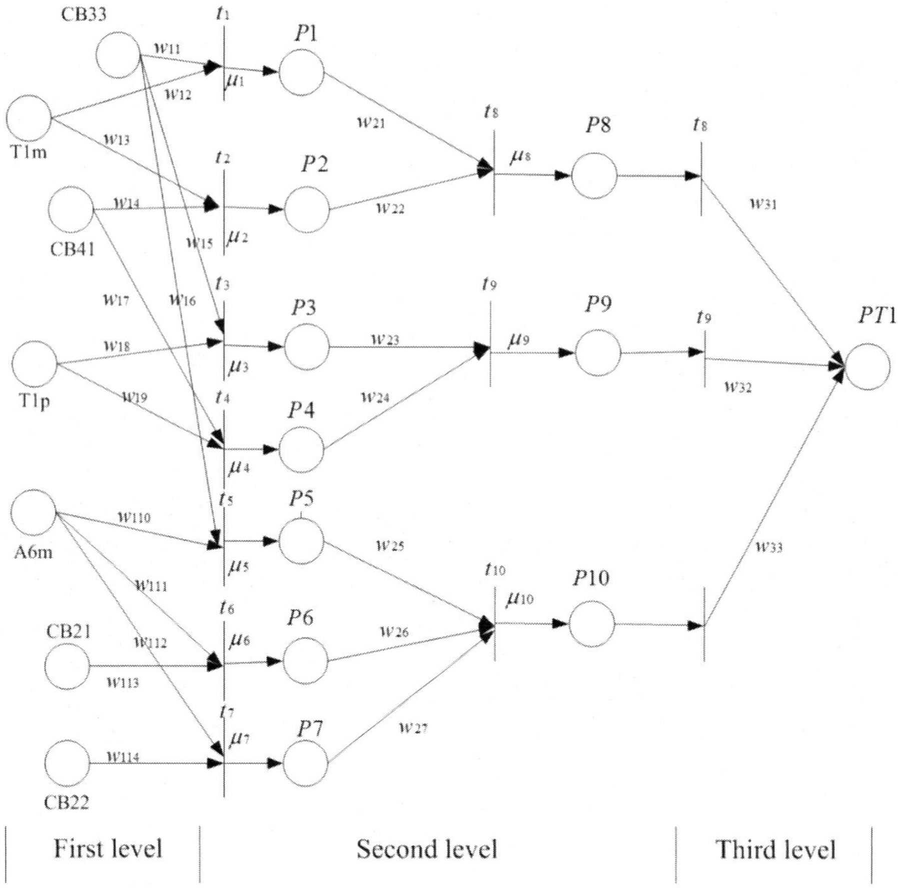

Figure 12. The FPN model of traction transformer T1.

Table 7. The operating time information of protection devices in the fault reasoning (by the end of December 2013)

Protection name	January 2009		June 2011		December 2013	
	elapsed time (hour)	reliability	elapsed time (hour)	reliability	elapsed time (hour)	reliability
protection of bus B1	17280	0.9922	38160	0.9128	59760	0.6938
protection of bus A6	17280	0.9922	7920	0.9993	20880	0.9860
protection of line L13	5040	0.9998	25920	0.9728	47520	0.8354
protection of transformer T1	17280	0.9922	38160	0.9128	59760	0.6938
protection of transformer T2	17280	0.9922	38160	0.9128	59760	0.6938

This TPSS was put into operation from January 2007; most of the equipment has been used a long time. The protection device of line L13 was replaced in June 2008, the protection device of bus A6 was replaced when it malfunctioned in July 2010, and the circuit breaker

CB42 was replaced in January 2013. The used equipment operating time information from Figure 11 to Figure 13 is shown in Table 7 and Table 8, the reliability in tables are calculated in (26) and (27). It can be seen that the same equipment has different credibility at different times and the credibility reduces as the operating time is extended.

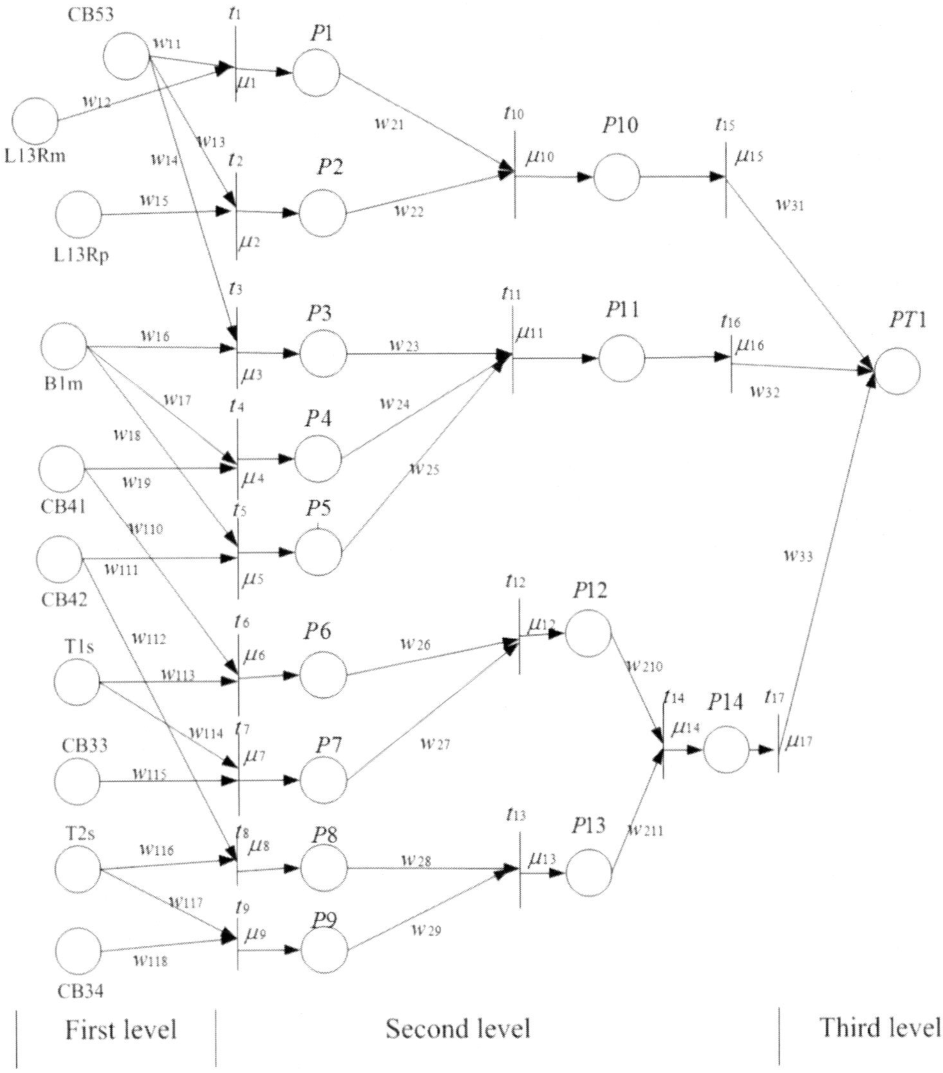

Figure 13. The FPN model of line L13.

Using the listed reliability in the tables above and according to (30), the weight coefficient of each part in Figure 11 is shown in Table 3-9. And the weight coefficient of each part in Figure 12 and Figure 13 is shown in Table 10 and Table 11, respectively.

Table 8. The operating time information of circuit breakers in the fault reasoning (by the end of December 2013)

Circuit breaker name	January 2009		June 2011		December 2013	
	number of action (times)	reliability	number of action (times)	reliability	number of action (times)	reliability
CB21	360	0.9995	780	0.9983	1143	0.9968
CB22	336	0.9996	673	0.9987	1276	0.9962
CB33	424	0.9994	840	0.9981	1366	0.9957
CB34	472	0.9992	957	0.9976	1407	0.9955
CB41	640	0.9988	1356	0.9957	1867	0.9928
CB42	680	0.9986	1421	0.9954	2018	0.9918
CB53	1360	0.9957	3243	0.9823	217	0.9998

Table 9. The weight of each part in the fault FPN model of bus B1

weight number	weight value			weight number	weight value			weight number	weight value		
	2009.1	2011.6	2013.12		2009.1	2011.6	2013.12		2009.1	2011.6	2013.12
w_{11}	0.5017	0.5217	0.5886	w_{17}	0.4983	0.4783	0.4114	w_{23}	0.3327	0.3303	0.3350
w_{12}	0.4983	0.4783	0.4114	w_{18}	0.4984	0.4783	0.4116	w_{24}	0.5000	0.5001	0.5002
w_{13}	0.4984	0.4783	0.4116	w_{19}	0.5017	0.5217	0.5886	w_{25}	0.5000	0.4999	0.4998
w_{14}	0.5016	0.5217	0.5884	w_{110}	0.5016	0.5217	0.5884	w_{31}	0.4970	0.4727	0.4096
w_{15}	0.4991	0.4817	0.4097	w_{21}	0.3337	0.3349	0.3327	w_{32}	0.5030	0.5273	0.5904
w_{16}	0.5009	0.5183	0.5903	w_{22}	0.3336	0.3348	0.3323				

Table 10. The weight of each part in the fault FPN model of transformer T1

weight number	weight value			weight number	weight value			weight number	weight value		
	2009.1	2011.6	2013.12		2009.1	2011.6	2013.12		2009.1	2011.6	2013.12
w_{11}	0.5018	0.5223	0.5893	w_{19}	0.4983	0.4783	0.4114	w_{23}	0.5002	0.5006	0.5007
w_{12}	0.4982	0.4777	0.4107	w_{110}	0.4982	0.5003	0.4975	w_{24}	0.4998	0.4994	0.4993
w_{13}	0.4983	0.4783	0.4114	w_{111}	0.4982	0.5003	0.4973	w_{25}	0.3333	0.3332	0.3332
w_{14}	0.5017	0.5217	0.5886	w_{112}	0.4981	0.5002	0.4974	w_{26}	0.3333	0.3333	0.3335
w_{15}	0.5018	0.5223	0.5893	w_{113}	0.5018	0.4997	0.5027	w_{27}	0.3334	0.3334	0.3333
w_{16}	0.5018	0.4997	0.5025	w_{114}	0.5019	0.4998	0.5026	w_{31}	0.3342	0.3126	0.2505
w_{17}	0.5017	0.5217	0.5886	w_{21}	0.5002	0.5006	0.5007	w_{32}	0.3342	0.3126	0.2505
w_{18}	0.4982	0.4777	0.4107	w_{22}	0.4998	0.4994	0.4993	w_{33}	0.3317	0.3748	0.4989

Table 11. The weight of each part in the fault FPN model of line L13

weight number	weight value			weight number	weight value			weight number	weight value		
	2009.1	2011.6	2013.12		2009.1	2011.6	2013.12		2009.1	2011.6	2013.12
w_{11}	0.4990	0.5024	0.5448	w_{112}	0.5016	0.5078	0.5884	w_{25}	0.3336	0.3348	0.3323
w_{12}	0.5010	0.4976	0.4552	w_{113}	0.4983	0.4783	0.4114	w_{26}	0.4998	0.4994	0.4993
w_{13}	0.4990	0.5024	0.5448	w_{114}	0.4982	0.4777	0.4107	w_{27}	0.5002	0.5006	0.5007
w_{14}	0.5009	0.5183	0.5903	w_{115}	0.5018	0.5223	0.5893	w_{28}	0.4998	0.4994	0.4991
w_{15}	0.5010	0.4976	0.4552	w_{116}	0.4984	0.4922	0.4116	w_{29}	0.5002	0.5006	0.5009
w_{16}	0.4991	0.4817	0.4097	w_{117}	0.4982	0.4917	0.4107	w_{210}	0.5001	0.4724	0.5003
w_{17}	0.4983	0.4783	0.4114	w_{118}	0.5018	0.5083	0.5893	w_{211}	0.4999	0.5276	0.4997
w_{18}	0.4984	0.4783	0.4116	w_{21}	0.5000	0.5000	0.5000	w_{31}	0.3387	0.3774	0.5569
w_{19}	0.5017	0.5217	0.5886	w_{22}	0.5000	0.5000	0.5000	w_{32}	0.3315	0.3061	0.2625
w_{110}	0.5017	0.5217	0.5886	w_{23}	0.3327	0.3303	0.3350	w_{33}	0.3298	0.3165	0.1806
w_{111}	0.5016	0.5217	0.5884	w_{24}	0.3337	0.3349	0.3327				

According to the data from Table 9 to Table 11 by the means of FPN, it can be seen that the weight of each place of information, in reasoning, is no longer set artificially and changes dynamically as the reliability relationship of equipment.

Using the weight of each part, which is calculated according to Table 9 thru Table 11, combined with the reliability of the protection devices and circuit breakers in Table 7 and Table 8, several typical fault conditions are diagnosed using this method and the results are shown in Table 12.

The methods of this section take into account the time characteristics of protection devices, operating reliability of circuit breakers, and the relationship between source reliability of FPN fault diagnosis and equipment operating time, so the credibility of the diagnosis is affected by equipment service time and is closer to the reality, so the diagnosis result is more convincing.

2. Reliability Prediction

If the interruption of HSR caused by TPSS faults takes place, enormous economic losses and serious social impacts will be brought in. Therefore, to infer, assess, and predict the health state and reliability of TPSS and to take preventive measures according to the prediction results in advance can help to eliminate faults in the embryonic stage and to reduce or even avoid the loss caused by TPSS faults.

The TPSS equipment arranged outdoors and the operating state is easily affected by the external environment, such as wind, rain, snow weather, lightning and fog. Further, HSR in China is almost arranged on the overhead viaduct leading to stronger influence of TPSS by various meteorological conditions. The heavy weather conditions have become an important factor contributing to faults of HSR TPSS. Prediction and assessment of the influence on TPSS according to the meteorological conditions is key to HSR scheduling management.

Table 12. The diagnosis results of the case

fault signal	diagnosis results/credibility		
	2009.1	2011.6	2013.12
B1m action; CB41, CB42, CB53, CB54 action	bus B1 fault, credibility 0.8535	bus B1 fault, credibility 0.8196	bus B1 fault, credibility 0.7469
T1m, A6m action; CB33, CB41, CB21 action	Transformer T1 fault, credibility 0.8538	Transformer T1 fault, credibility 0.8564	Transformer T1 fault, credibility 0.8498
L13Rm action; CB53, CB54 action	line L13 fault, credibility 0.8555	line L13 fault, credibility 0.8381	line L13 fault, credibility 0.7930

2.1. Analysis of environmental meteorological factors affecting TPSS

2.1.1. The distribution regulations of TPSS faults

The railway transportation has the character of changing with time, and the external environment of TPSS operation also has the character of changing with time. As a result, the fault distribution of TPSS also has certain regularity with time. Studying the distribution of the TPSS fault regarding time can help to formulate preventive measures and corresponding countermeasures targeted to focus on periods, in order to prevent and reduce the faults effectively.

The example of a specific TPSS section in nearly three years is taken. The time variable regulations of its faults are analyzed. Statistics data include all traction substations and all catenary faults of this TPSS.

1) The monthly regulations of TPSS faults

It is obvious that weather changes seasonally and cyclically while the faults number is time variable. Figure 14 shows the monthly fault curve of the TPSS in nearly three months according to the order of the four seasons.

As it can be seen in Figure 14, the fault distribution of TPSS has obvious seasonal characteristics, and summer and winter are the peak season of faults. Among them August was the peak month of faults in 2010, which had 24 faults. July was the peak of faults in 2011 with 38 faults. August was the peak month of faults in 2012 with up to 45 faults. The peak of faults in the most recent three years all appeared in summer.

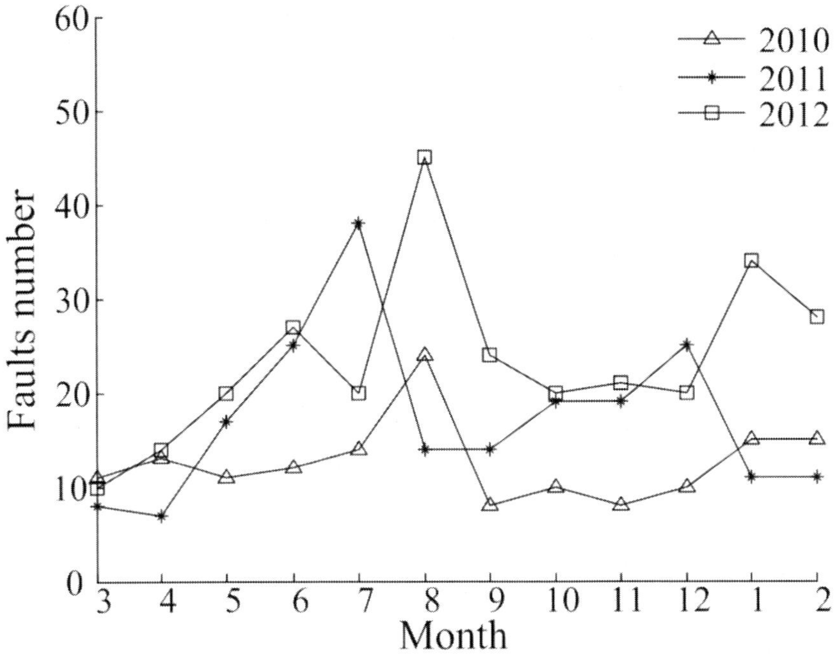

Figure 14. Monthly faults curve.

In addition, it can be seen from Figure 14 that the number of TPSS faults increased year by year. The number of faults in 2010, 2011 and 2012 is 151, 208 and 283, respectively. The number of faults increased significantly caused by the gradually accumulated operating damage.

2) The daily regulations of TPSS faults

Various frequency and intensity of High-Speed train operations in one day led to the various fault distributions of TPSS in that day. Figure 15 and Figure 16 show the daily fault curve of TPSS in different seasons.

From Figure 15, the peak of faults is relatively concentrated in summer and winter. The fault peak appeared at 15:00 and 16:00 in summer, which is the time reached for the highest temperature. Otherwise, the peak of faults appeared in 9:00-11:00 and 13:00-15:00.

From Figure 16, the daily fault curve in spring and autumn is average. The number of faults distributed in a longer period of time and a smaller fault peak appeared in 8:00-12:00 and 13:00-18:00.

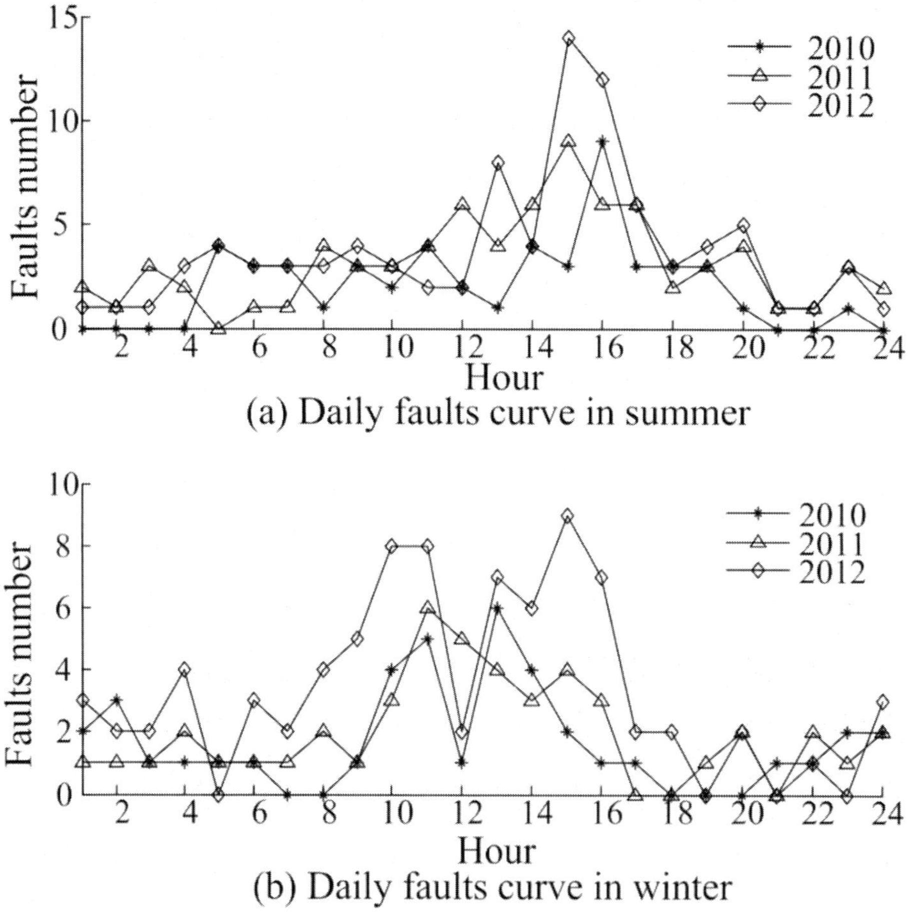

Figure 15. Daily faults curve in summer and winter.

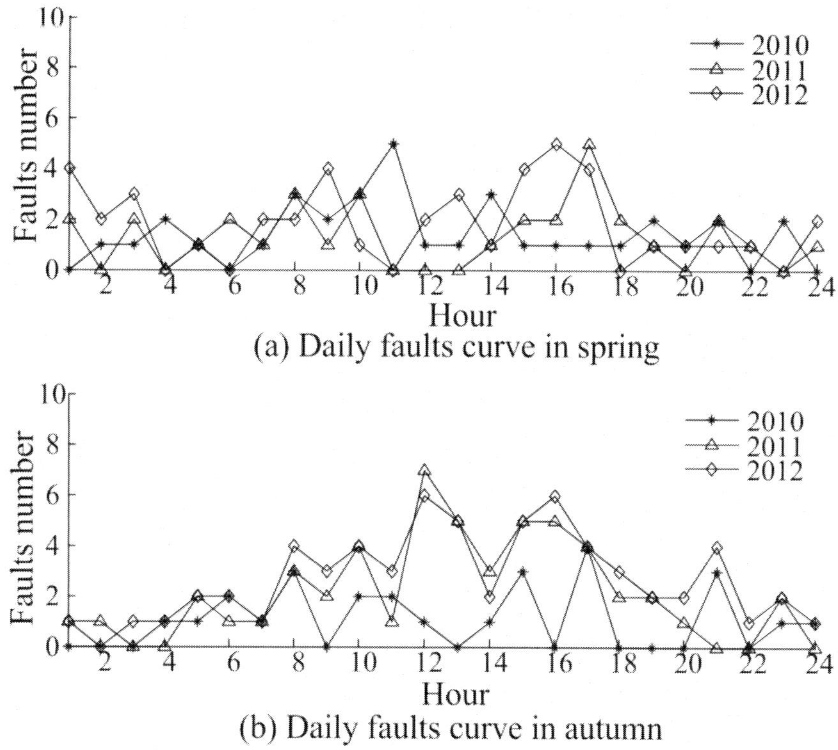

Figure 16. Daily faults curve in spring and autumn.

2.1.2. Analysis of faults and meteorological factors relation

TPSS operating status is strongly affected by a variety of meteorological factors. Theoretically, the greater the number of meteorological factors, the better the analysis result will be. But in fact several important factors are enough, as follows.

1) The wind. The influence of wind on the TPSS is mainly reflected in the overhead catenary system. Under the effect of wind, the internal forces of catenary lines may exceed the allowable value, causing pull cut of the wire; Catenary vibrations caused by wind lead to structural fatigue; Catenary wield caused by wind will disturb the normal current collection of the pantograph. Interaction strength of wind is proportional to the square of speed. Therefore, when the speed of wind reaches a certain level, the influence is more prominent on catenary equipment.
2) The rain. The influence of rain is mainly reflected in the weakened insulation performance of TPSS equipment. Rain increases air humidity and weakens the insulation performance of equipment. In addition, pollution flashover happens when dirt on the insulator surface is damp. Water flow on the equipment surface may lead to short circuits.
3) The temperature. Temperature will affect the service conditions of the catenary line. Temperature changes will cause the catenary line to sag, which changes the tension of the catenary line. It finally has an influence on the geometric construction of the catenary.

4) Lightning. Lightning is one of the most dangerous risk factors in TPSS. Lightning on a catenary system will produce overvoltage, insulator damage, and even line trips. Furthermore, overvoltage will invade the traction substation and lead to equipment damage.
5) Average relative humidity. The average relative humidity is similar to the influence by rain, and mostly affects the insulation performance of TPSS equipment.

Figure 17 and Figure 18 show the example section of TPSS fault statistics and the corresponding meteorological factors in the area in the most recent three years. These figures show the changes of wind speed, temperature, rain precipitation, lightning times and average relative humidity. From these figures, the lightning times have the strongest influence on TPSS faults. The peak of lightning times is approximately coincidental to the peak of faults number in each year. The peak of faults and the peak of temperatures also have a high consistency, both appearing in the annual summer. The faults number also has some relevance with wind speed, rain precipitation and relative humidity.

In Figure 17, relativities between faults and lightning show that the peak of lightning times approximately coincide with the peak of faults number in the three years. Both in 2010 and 2012, lightning times peaked in August while the number of faults also peaked in August. Similarly, lightning times peaked in July while the number of faults also peaked in July, 2011.

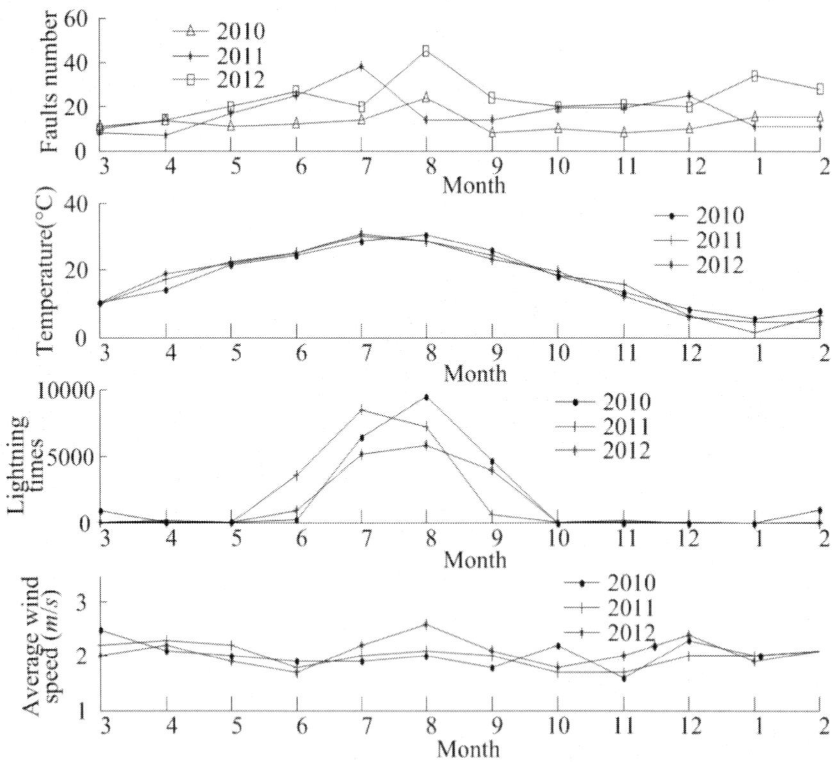

Figure 17. Monthly faults curve and temperature, lightning, and wind curve of TPSS in the last 3 years.

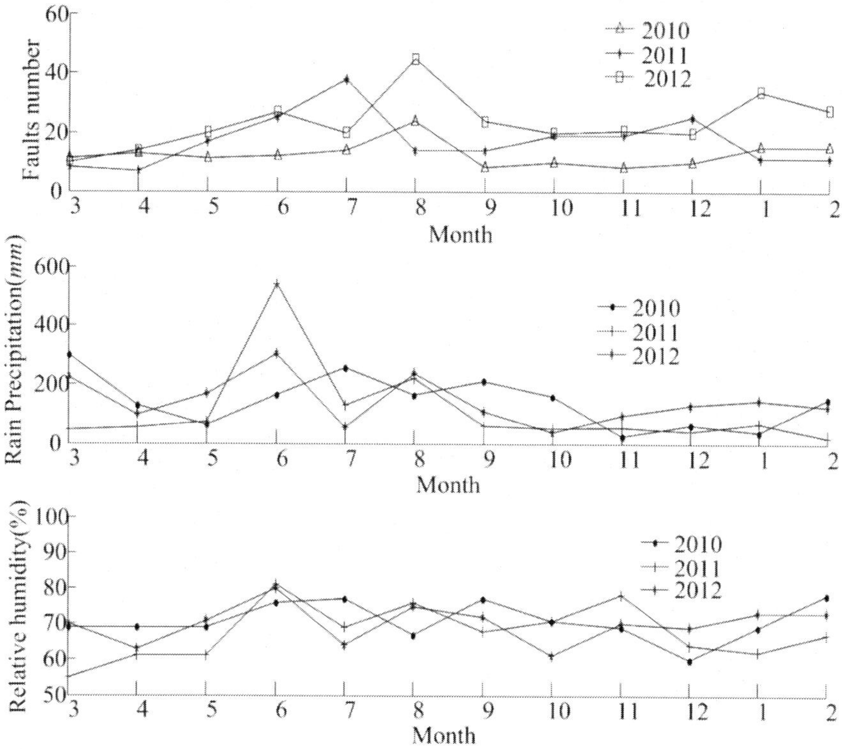

Figure 18. Monthly faults curve and rain and relative humidity curve of TPSS in the last 3 years.

From these figures, the relationship between the number of faults and the temperature is obvious. The peak of faults and the peak of temperature is also highly consistent. It shows that the highest value of average temperature and the annual faults peak appeared in August 2010, and then appeared in July in 2011. In 2012, the highest value of average temperature appeared in July, followed by the annual faults peak that appeared in August.

In Figure 18, faults number of TPSS show certain relevance with monthly rain precipitation and monthly relative humidity in this area, but the relationship is not obvious.

It can be seen by the above analysis that there are some relevance and association relationships between the TPSS faults and various meteorological factors, although the degree of these relationships differs from factor to factor.

Make variables x the average monthly statistical value of meteorological factors, and variables y as the monthly fault numbers of TPSS. The correlation coefficient is:

$$r = \frac{\sum_{i=1}^{12}(x_i - \overline{x})(y_i - \overline{y})}{\sqrt{\sum_{i=1}^{12}(x_i - \overline{x})^2 \cdot \sum_{i=1}^{12}(y_i - \overline{y})^2}} \tag{32}$$

The correlation coefficient between the faults number and relevant meteorological factors in the past three years are obtained and shown in Table 13.

Table 13. The correlation coefficient between the faults number and relevant meteorological factors in the past three years

items	correlation coefficient *r*		
	2010	2011	2012
Temperature	0.3816	0.3990	0.4577
Lightning times	0.5945	0.5819	0.6438
Wind speed	0.4777	0.4232	0.4585
Relative humidity	0.3498	0.3672	0.3043
Rain precipitation	0.3391	0.1998	0.2501

The correlation coefficient can reflect the correlation degree between the mutual variables. From Table 13, there are positive correlation relationships between the faults number and relevant meteorological factors. The correlation coefficient can be divided into 3 according to the correlation degree: $|r| < 0.4$ is a low linear correlation, $0.4 \le |r| < 0.7$ is a significant correlation, $0.7 \le |r| < 1$ is a highly linear relationship. Therefore, the relationship between lightning times, wind speed and the faults number of the TPSS has a significant correlation. The relationship between the temperature and the faults number is significant in 2012, but has a low linear correlation in 2010 and 2011. The relationship between rain precipitation, relative humidity and the faults number has low linear correlation.

The correlation coefficient shows the correlation degree of the faults number and meteorological factors. But practically obtaining the correlation coefficient of them is not enough. Therefore, it is necessary to do further analysis to clarify the specific relationships.

2.2. Regression analysis and prediction of TPSS faults

2.2.1. Variable election of FMRM

There are many factors influencing the service status of TPSS. Some of the factors are significant while the others are relatively mild. How to determine more important factors that affect the TPSS faults from a lot of factors, such as the the independent variable of the meteorological analysis model, is very important for the determination of the whole analysis model.

Use the stepwise regression method to choose the variables of the FMRM. The basic idea of the stepwise regression method is as follows:

1) Use the fault frequency of the TPSS y and the meteorological factors $x_i \ (i = 1, 2, \cdots 5)$ to take regression analysis respectively and the value of its significance test F is obtained. If one of the biggest F values under a given level of significance is significant, the corresponding meteorological factors is selected.
2) For the rest of the $m-1$ variables, reuse Partial-F test to select a variable into the model, the method of Partial-F statistic is

$$F = \frac{SSR(X_j | X_i)/k}{MSE(X_i, X_j)} \sim F(1, n-2) \tag{33}$$

where $SSR(X_j|X_i)=SSR(X_i,X_j)-SSR(X_i)$ explained the contribution of the added new variable to the dependent variable under the condition of existing model variables. $SSR=\sum(y_i-\bar{y})^2$ means regression quadratic sum. $MSE(X_i,X_j)=SSE(X_i,X_j)/(n-k-1)$, $SSE=\sum(y_i-y_i)^2$ means the residuals quadratic sum.

3) The significance test is made for each variable in the model. If the test is not significant, get rid of the variable, repeat 2) and 3); Or retain the variable, repeat 2) and 3) until all the selected variables according to a given level of significance no longer change.

Stepwise regression selected is a self-renewing process. The unimportant TPSS faults and meteorological factors ultimately will not be introduced, even if they were introduced at the beginning, they will be removed last. Until no new variable is introduced, no variables are eliminated and the stepwise regression completes.

2.2.2. Establishment of FMRM

After determining the meteorological factors that influence the service status of TPSS, it is necessary to clarify the relationship between the faults number and the meteorological factors so that the actual work can determine the degree of changes in the service status of TPSS according to the changes of meteorological conditions. The core problem is the estimation of the parameters and its assessment.

The TPSS faults statistical sample is expressed in Y and the sample observations of meteorological factors is expressed in X. The relationship between Y and X could be expressed by a conditional expectation equation as:

$$Y=E(Y|X)+\varepsilon \tag{34}$$

where $E(Y|X)$ are the average conditions of variable Y on variable X (can be a variable, or can be composed of multiple variables vector). ε is the random error.

Use $B=[\beta_0,\beta_1,\beta_2,\cdots,\beta_n]^T$ to stand for the regression coefficient so the regression equation can be represented as: $Y=XB+\varepsilon$

As a result, to establish the regression model, the key and core is the assessment of the regression coefficient B. Use the method of least squares to assess B as :

$$\hat{B}=(X'X)^{-1}X'Y \tag{35}$$

After the regression equation is assessed, regression models need to be diagnosed using the goodness of fit index R^2 and residual analysis to reflect the fitting degree of the equation. Goodness of fit index as:

$$R^2 = \frac{SSR}{SST} = \frac{SSR}{SSR + SSE} \tag{36}$$

where the regression quadratic sum SSR reflects the independent variable's contribution to the total deviation. The residuals quadratic sum SSE reflects the error variable's contribution to the total deviation. The greater the R^2 shows that the fit is better.

If the regression model of the sample data fits well, the assessment of error variable $\hat{\varepsilon}$ should be satisfied with the properties of zero mean, same variance, irrelevant and normality. Make $r_i = \hat{\varepsilon}_i = y_i - \hat{y}_i$ as residual and $e_i = \frac{r_i}{\sigma} = \frac{r_i}{\sqrt{MSE}} \sim N(0,1)$ as standardized residuals. That is to say, the assessment of the standardized residuals should approximately obey the standard normal distribution.

After the regression model successfully passed all the above tests, the prediction of the dependent variable according to the actual situation of the independent variables, the change can be executed. The regression prediction is based on the known meteorological conditions X_0, to predict the possible values $E(Y_0|X_0)$ for the TPSS faults number.

Combined with the regional weather forecast, the prediction of the average number of TPSS faults according to meteorological reason can be carried out. The $100(1-\alpha)\%$ confidence interval of the average value $E(Y_0|X_0)$ is estimated as:

$$(\hat{Y}_0 - t_{\alpha/2}(n-p-1)s(\hat{Y}_0),\ \hat{Y}_0 + t_{\alpha/2}(n-p-1)s(\hat{Y}_0)) \tag{37}$$

where $s(\hat{Y}_0)$ is the assessment square root of the variance $\hat{Y}_0$:

$$s(\hat{Y}_0) = \hat{\sigma}\sqrt{X_0^{'}(X'X)^{-1}X_0} \tag{38}$$

2.2.3. Case study

Choose the last three years (2010-2012) of the TPSS faults as the research case and analyze the relationship between the faults number and meteorological conditions, and the FMRM is established. The fault statistics data of the TPSS includes all the traction substations and catenary faults under the jurisdiction. The meteorological data comes from the statistics in the region of the local weather department.

Coordinate the faults situation in the last three years and get the TPSS faults number, per month, in the last three years, respectively, named as dependent variable y. Coordinate corresponding meteorological conditions in the region and obtain the lightning times per month x_l, average monthly temperature x_t, average monthly wind speed x_w, average monthly rain precipitation x_r, average monthly relative humidity x_h for three years in the region, respectively, noted for the independent variables $x = (x_t, x_l, x_r, x_w, x_h)$. The change of variable y and x is shown in Figure 17 and Figure 18.

First of all, without variable election, directly analyze the faults number and five variables regression of meteorological factors and get the relationship between the faults number y and the meteorological factors:

$$y = 1.2413 + 0.0137x_t + 3.7515x_l + 0.0014x_r + 0.3320x_w + 0.0112x_h$$

The residual leverage figure after the regression is shown in Figure 19. Figure 19 shows that residual uniform distribution is near the zero line. But high leverage points appear in 17th data, which means that 17th data is the abnormal point and should be removed when fitting. The test parameters of returned value $R^2 = 0.6914$ show that the model fit is not very good. Test value $F = 1.9673 > p$ meets the requirements. However, the value associated with significant probability $p = 0.1824 > 0.05$ and this shows that some variables in the regression equation can be eliminated.

After stepwise regression variable selection, the regression equation is obtained as follows:

$$y = 1.5349 + 0.3314x_w + 3.7229x_l$$

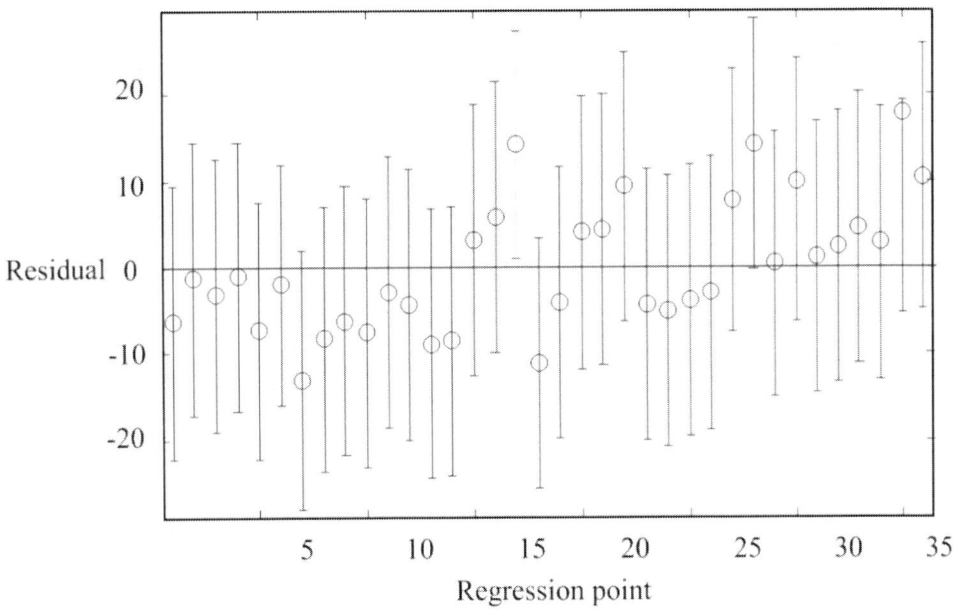

Figure 19. The residual leverage figure after regression analysis.

At this time, regression equations retain the original variables x_l, x_w respectively on lightning times and average wind speed. The assessment parameters of the model are R^2 =0.8493 and test value F =7.4761> p. The value associated with significant probability p =0.0274<0.05. The above indexes show that the regression effect is ideal. Regression results show that the faults number of TPSS, and lightning and wind speed are closely related. This is consistent according to the actual situation experience judgment.

Two methods of fitting assessment parameters are shown in Table 14. In the table RMSE (root mean square error) shows mean square error of fitting. The smaller the value, the smaller the deviation between the fitting data and the real value, and the fitting effect is better. It could be seen from the data, after the variables selected, regression is better, the condition of the goodness of fit is improved further and the root mean square error is reduced further, which shows that the fitting result is more ideal.

By the model, after stepwise regression variable selection to forecast the faults number in the TPSS for March to May in 2013, the forecasting situation is as shown in Table 15.

It can be seen from Table 15 that using the regression model to predict the accident of TPSS has a certain feasibility and its 95% confidence interval prediction can roughly reflect the faults number. Thus, the service status of the TPSS is influenced by many factors except for not only the meteorological conditions, but also some external emergency. Prediction according to the weather condition tends to be more conservative than the actual situation, but the prediction results can offer practical reference for daily operation and management of TPSS.

As the sample data of TPSS faults accumulate gradually, the result of regression analysis will tend to be more accurate and prediction results will be closer to the actual.

Table 14. The comparison of the assessment parameters of the two kinds of regression model

methods	Comparison			
	R^2	F	P	RMSE
Without variable selection	0.6914	1.9673	0.1824	12.8027
After stepwise regression variable selection	0.8493	7.4761	0.0274	7.5549

Table 15. Regression model used in faults prediction in TPSS

timeline	95% confidence interval prediction	The actual value	Predict range
2013.3	(9.8192,13.0380)	11	Within the scope of prediction
2013.4	(6.8615,11.0947)	13	beyond the scope of prediction
2013.5	(6.3742,10.4036)	11	beyond the scope of prediction

2.3. Fault probability distribution statistical and prediction in TPSS

2.3.1. The chain structure and reliability distribution of TPSS

The TPSS is composed of a traction substation and catenary system: the traction substation obtains energy from an external power supply. Energy through transformer's step-down and commutation to High-Speed train through feeder and catenary. And the current returns to the substation through the rail at last, which composes a complete circuit. The traction substation is composed of an incoming circuit breaker, disconnector, traction transformer, bus, feeder breaker and feeder, etc. Different anchoring sections connect and form the catenary system. So, based on the flow direction of power, TPSS is a chain structure circuit combined with many power equipment supplies and elements; the typical chain structure of TPSS is shown in Figure 20.

For the typical chain structure of TPSS, the fault of any link will cause the fault of the whole TPSS and influence the operation of TPSS. In the series system, only when all equipment in system is fine, can the system be in good condition. Therefore, reliability of TPSS is decided by reliability of the weakest link in TPSS.

Suppose the equipment of TPSS is represented by S_i $(i = 1, 2, \cdots, n)$, each equipment's normal condition is shown as A_i $(i = 1, 2, \cdots, n)$, and TPSS's normal condition is shown as A_s. Then the following equation is set up:

$$A_s = A_1 \cap A_2 \cap \cdots \cap A_n$$

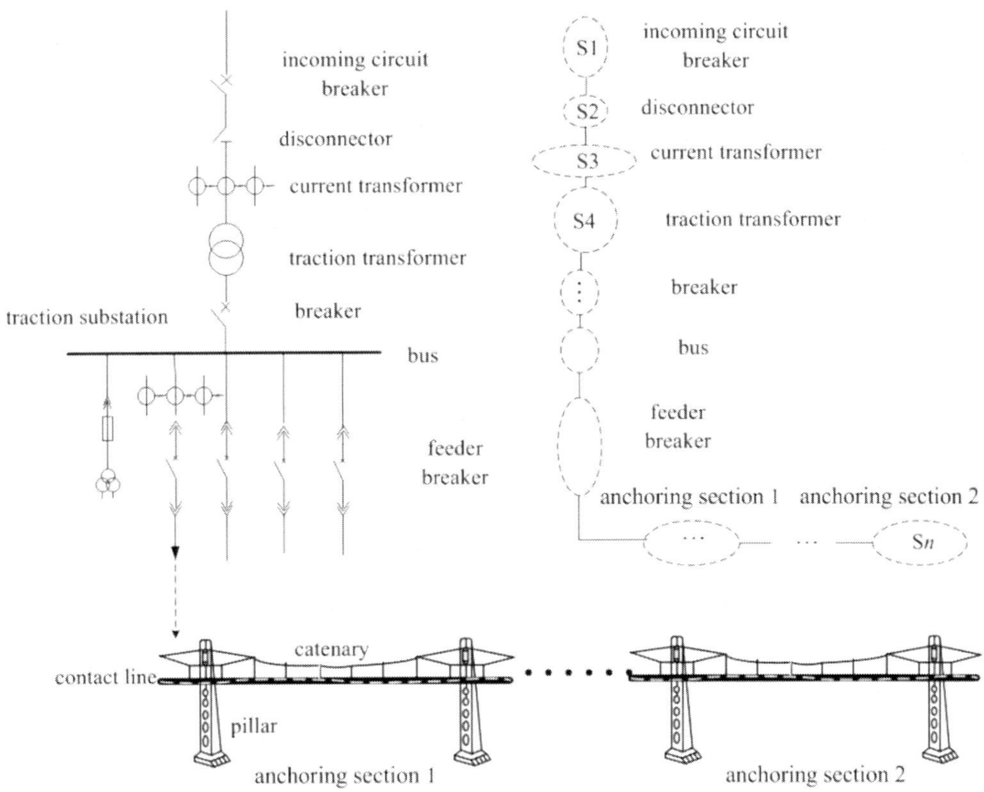

Figure 20. Schematic diagram of typical chain structure of TPSS.

And the reliability of TPSS is

$$R_s = P(A_s) = P(A_1 \cap A_2 \cap \cdots \cap A_n)$$

where R_s is the reliability of TPSS. $P(A)$ is the probability of event, A_s means every piece of equipment in TPSS is normal. A_i means the No. i equipment is in normal condition.

Each equipment's service condition is independent from each other, thus:

$$R_s = P(A_1)P(A_2)\cdots P(A_n) = \prod_{i=1}^{n} P(A_i) \tag{39}$$

For this kind of independent chain structure, Weibull has given the strength (life) probability distribution function as:

$$F(t) = 1 - \exp[-(\frac{t - t_0}{\eta})^m], \quad t_0 \le t < \infty \tag{40}$$

The reliability function is:

$$R(t) = \exp[-(\frac{t - t_0}{\eta})^m] \tag{41}$$

The failure rate function is:

$$\lambda(t) = \frac{m(t - t_0)^{m-1}}{\eta^m} \tag{42}$$

Where, t is the time. t_0 is non-failure operation time. m is the shape parameter, which reflects the changing trend of fault probability. η is the scale parameter, which is relevant to the minimum fault probability of the system.

As a typical chain structure, there is a shortest operation time t_0. TPSS is in normal and efficient operation status during t_0. When operation time $t < t_0$, the system would not go wrong. In other words:

$$F(t) = P(t \le t_0) = 0 \tag{43}$$

Since there are many links in TPSS, the random fault of each link has a remarkable effect on the system. Normal operation time t_0 is assumed to be 0 for simplicity, and the reliability function of TPSS can be simplified as Weibull distribution with two parameters.

The key to confirm Weibull distribution is to estimate the distribution parameter m and η. The least square method is the most common method to estimate parameters. The principle is shown as (17) to (19).

Failure distribution function $F(t)$ is very important in the least square estimation, generally using a middle rank estimation to obtain the function:

$$\hat{F}_i = \frac{i-0.3}{n+0.4}$$

Since the fault statistics in TPSS are mostly incomplete life cycle and the system is in operation when statistics are finished, the obtained fault statistics are just a stopping experiment to get incomplete samples. So using middle rank method would increase the error. To improve the accuracy of failure distribution function, using fault sample data and suspension sample data, estimate all possible average ranks. Then insert the rank in the middle rank formula to obtain the failure distribution function. This is the basic principle of the average rank method.

$$\Delta A = \frac{n+1-A_{k-1}}{n-i+2} \tag{44}$$

$$A_k = A_{k-1} + \Delta A_{k-1} \tag{45}$$

$$F(t_k) = \frac{A_k - 0.3}{n+0.4} \tag{46}$$

$$R(t_k) = 1 - F(t_k) \tag{47}$$

where A_k is the average rank of fault data. k is sequence number of fault data. A_{k-1} is the average rank of previous fault data. ΔA_k is the average rank increment. i is fault data's sequence number, which is arranged by time. t_k is runtime when the system is in No. k fault.

According to the fault data of TPSS after fitting parameters of Weibull distribution, the law of failure distribution, which is matched with fault data analysis can be figured out. It's reasonable to use the law of failure distribution to evaluate and predict the reliability of TPSS in the future.

2.3.2. Weibull distribution statistics of TPSS's reliability and the case study

Based on the fault statistics and data of a traction substation in the JiaoJi line, analyze, assess and predict the reliability of TPSS in this line.

Fault records are shown in Table 16 from June 2005.

Table 16. Fault time records of a traction substation (before December 2013)

Number	Fault time	Runtime (hours)	Status label	Number	Fault time	Runtime (hours)	Status label
1	2008-02-25	23976	F1	8	2011-8-25	54624	F8
2	2008-12-16	31056	F2	9	2011-12-22	57480	F9
3	2009-01-28	32088	F3	10	2012-4-7	60048	F10
4	2010-05-13	43368	F4	11	2012-8-19	63264	F11
5	2010-07-30	45240	F5	12	2013-08-29	72664	F12
6	2011-1-20	49416	F6	13	—		
7	2011-5-15	52176	F7				

According to the records, figure out the fault distribution function $F(t)$ by (44):

$$A_2 = A_1 + \Delta A_1 = A_1 + \frac{n+1-A_1}{n-i+2} = 1+\frac{8+1-1}{8-2+2}=2$$

$$F_2 = \frac{A_2 - 0.3}{8+0.4} = 0.20238095$$

In the same way, obtain the rank of every point and corresponding $F(t)$. The results are shown in Table 17.

Table 17. Results of the average rank method

i	t_i	ΔA_k	A_k	$F(t_k)$	$R(t)$	$\ln t_i$	$\ln[-\ln(R(t))]$
1	23976	1	1	0.0833	0.9166	10.0848	-2.4417
2	31056	1	2	0.2024	0.7976	10.3435	-1.4867
3	32088	1	3	0.3214	0.6786	10.3762	-0.9474
4	43368	1	4	0.4405	0.5595	10.6775	-0.5436
5	45240	1	5	0.5595	0.4405	10.7197	-0.1986
6	49416	1	6	0.2244	0.7756	10.6775	-1.3699
7	52176	1	7	0.2638	0.7362	10.7197	-1.1834
8	54624	1	8	0.3031	0.6969	10.808	-1.0184
9	57480	1	9	0.3425	0.6575	10.9082	-0.8691
10	60048	1	10	0.3819	0.6181	10.9592	-0.7317
11	63264	1	11	0.4213	0.5787	11.0551	-0.6035
12	72264	1	12	0.4606	0.5394	11.1881	-0.4823

Suppose the ordinate and abscissa is $\ln[-\ln(R(t))]$ and $\ln t$. Use the least square method to fit the data and obtain the fitting curve that has the smallest vertical residual square sum as Figure 21. Simultaneously, suppose $\ln[-\ln(R(t))]$ and $\ln t$ as abscissa and ordinate. Use the least square method to fit the data and obtain a fitting curve that has the smallest level residual square sum as Figure 22. The parameters figured by the two methods above are shown in Table 18.

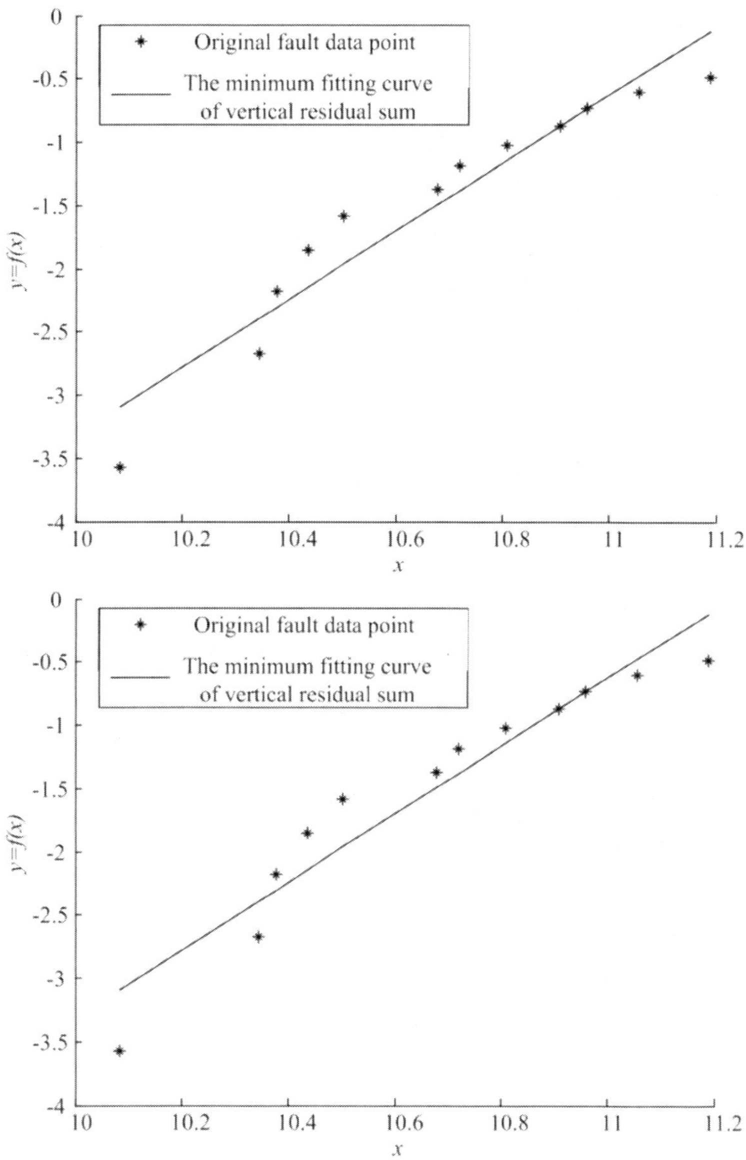

Figure 21. The minimum fitting curve of vertical residual sum.

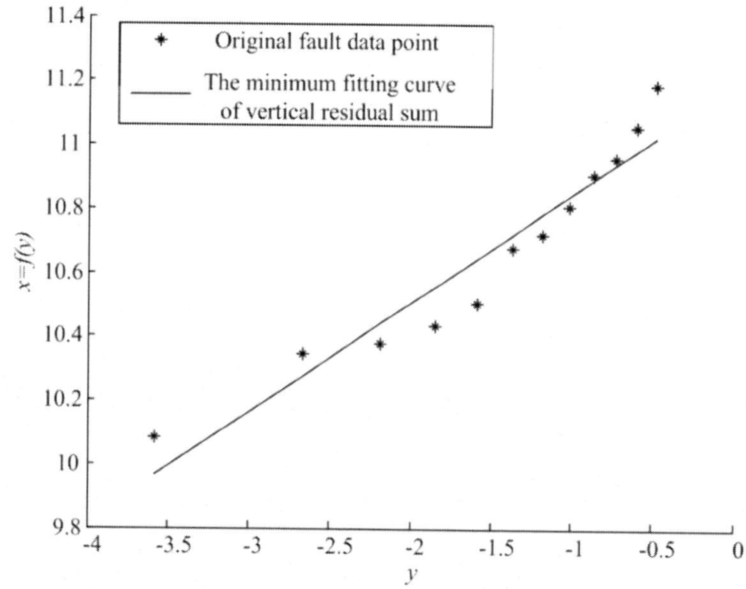

Figure 22. The minimum fitting curve of level residual sum.

Table 18. Parameters of two methods

method	m_{yx}	m_{xy}	η_{yx}	η_{xy}	RSS_{yx}	RSS_{xy}
Minimum vertical residual sum	2.6938		75515.1441		0.8806	
Minimum level residual sum		2.9344		72113.7491		0.3132

According to Table 18, the level residual, which using the method of the minimum level residual square sum is RSS_{xy} =0.3132, which is smaller than vertical residual RSS_{yx} =0.8806. So the result, which using the method of the minimum level residual square sum is more reasonable, is m_{xy} =2.9344, η_{xy} =72113.7491, which could more accurately reflect the actual law in the system.

The fault distribution function of the TPSS is:

$$F(t) = 1 - \exp[-(\frac{t}{72113.7491})^{2.9344}]$$

Distribution density function is:

$$f(t) = 1.6299 \times 10^{-14} \times t^{1.9344} \exp[-(\frac{t}{72113.7491})^{2.9344}]$$

Reliability function is:

$$R(t) = 1 - F(t) = \exp[-(\frac{t}{72113.7491})^{2.9344}]$$

Failure rate function is:

$$\lambda(t) = 1.6299 \times 10^{-14} t^{1.9344}$$

And the reliability function is shown in Figure 23. The shape parameter is m =2.9344>1, which shows that the traction substation has been in wastage. The substation was first placed in service in June 2005 and was working for 8 years by 2013. It has been in multiple fault phase, the reliability is integrally in a lower condition.

Without the consideration of repairing influence, the system's reliability $R(57744h) = 59.40\%$ Jan. 1, 2012, which belongs to the phase of multiple fault under the high pressure of maintenance.

Actually, the equipment in TPSS should be timely replaced or repaired when it goes wrong. As a consequence, the reliability of TPSS is higher than 59.4% in practice.

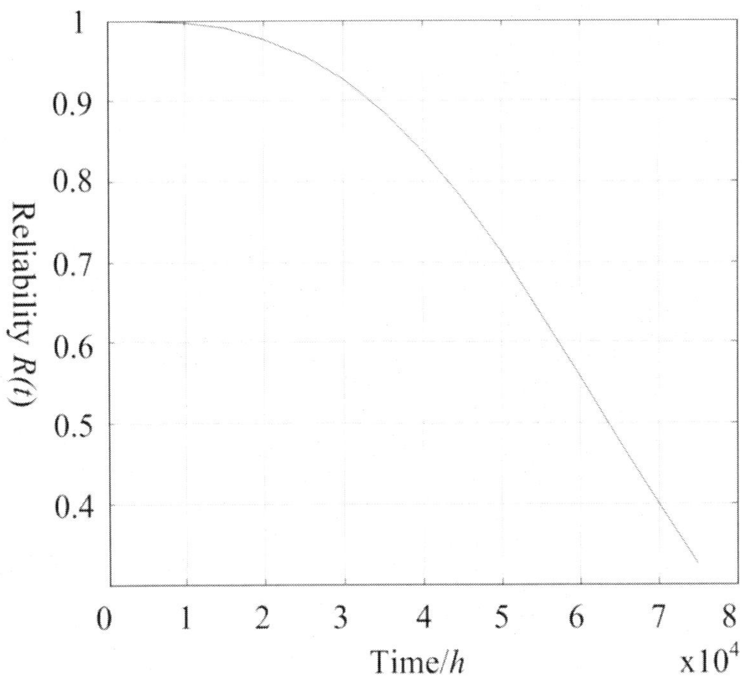

Figure 23. The reliability curve of the TPSS.

For the specific equipment (such as traction transformer, breaker, a section of contact line, etc.) in TPSS, using this method can also evaluate and predict reliability of the equipment if it is possible to obtain its detail service and fault records and predict the remaining life by analyzing the manufactured information of the equipment.

CONCLUSION

This chapter focus on two issues of HSR TPSS – fault diagnosis and reliability prediction – based on CBR techniques.

In terms of fault diagnosis, Weibull reliability curves of the protection device and circuit breaker are presented. Parameter estimation methods for reliability function are analyzed. The initial degrees of confidence of protection device and circuit breaker placement of FPN for TPSS are described by their reliability indices while the weight of each place is established. A section of TPSS is proposed as a case to test the method and the following conclusions can be obtained:

(1) Weibull distribution can well describe the reliability of protection devices and circuit breakers. For most equipment in the aging period, the parameter estimation method based on the minimum level residual sum of squares is more accurate.
(2) The initial degrees of confidence of FPN places, described by the reliability function, can reduce the subjectivity and conform better to an actual situation.
(3) The average rate of correct action of protection device and circuit breaker is replaced by reliability function. Therefore, the rule of fault diagnosis under uncertain information can be reflected.
(4) Information provided by the equipment with higher reliability has a greater weight when reasoning. It makes higher credibility of reasoning results.
(5) The confidence of fault diagnosis results are related to equipment service time. It makes the diagnosis closer to reality.

As for reliability prediction, meteorological factors affecting the operation state of TPSS are analyzed. Future fault conditions and reliability of TPSS are inferred, estimated and predicted from the perspective of statistics and probabilities. Relations between fault and meteorological factors are analyzed and FMRM is established to predict faults in TPSS. The fault time distribution of TPSS is described by the probability distribution method and the following conclusions are obtained:

(1) Historical fault data contain abundant information that can be mined by regression analysis and probability statistics to provide the reference for future status prediction.
(2) In the process of establishing FMRM, regression analysis after choosing the variables is much more effective than analysis without choosing the variables.
(3) TPSS is a typical system with chain structure. So its reliability can be described by Weibull distribution.
(4) Fault of TPSS correlates positively with meteorological factors, especially the lightning times. This suggests that lightning is the biggest threat to the safety and reliability of TPSS.

REFERENCES

Alberto, A., Arias, E., Rojo, T., et al., (2007). Numerical simulation of the static stiffness problem on a catenary by using treads. *Computation in Modern Science and Engineering, 2*, 1141-1145.

Alefeld, G. & Mayer, G. (2000). Interval Analysis: Theory and Application. *Journal of Computational and Applied Mathematics., 121*, 421-464.

Andrea, Collina. & Stefano, Bruni. (2002). Numerical simulation of pantograph-overhead equipment interaction. *Vehicle System Dynamics., 38*(4), 261-291.

Dong, L., Xiao, D., Liang, Y., et al., (2008). Rough set and fuzzy wavelet neural network integrated with least square weighted fusion algorithm based fault diagnosis research for power transformers. *Electric Power Systems Research., 78*(1), 129-136.

Gholamhossein, Yari. & Alireza, Chaji. (2012). Determination of Ordered Weighted Averaging Operator Weights Based on the M-Entropy Measures. *International Journal of Intelligent Systems, 27*(12), 1020-1033.

Mirrasoul, J., Mousavi, Karen L. Butler-Purry, (2010). Detecting incipient faults via numerical modeling and statistical change detection. *IEEE transactions on power delivery, 25*(3), 1275-1283.

Morais, D. & Rolim, J. (2006). A hybrid tool for detection of incipient faults in transformers based on the dissolved gas analysis of insulating oil. *IEEE transactions on Power Delivery, 21*(2), 673-680.

Patha, Kj., Jiang, Y., Honavar, V., et al., (2006). Condition data aggregation with application to failure rate calculation of power transformers. *Proceedings of the 9th International Conference on Probabilistic Methods Applied to Power Systems, 127.*

Qiu, Xiao-long. & Li, Ran. (2007). Fault Diagnosis of Transmission Network using Fuzzy Petri Nets. *Proceedings of Power Tech, IEEE*, 1802-1806.

Saghafi, A., Mirhabibi, A. R. & Yari, G. H. (2009) Improved linear regression method for estimating Weibull parameters. *Theoretical and Applied Fracture Mechanics, 52*, 180-182.

Weibull, W. (1951). A Statistical distribution function of wide applicability. *Journal of Applied Mechanics., 18*, 293-297.

Xiaoou, Li., WEN, Yu. & Felipe, Lara-Rosano. (2000). Dynamic knowledge inference and learning under adaptive fuzzy net framework. *IEEE transactions on systems, man, and cybernetics-Part C: applications and reviews, 30*(4), 442-450.

Xu, Luo. & Mladen, Kezunovic. (2008). Implementing fuzzy reasoning Petri-Nets for fault section estimation. *IEEE transactions on power delivery, 23*(2), 676-685.

Yilu, Zhang, Gary W, Gantt., Jr. et al., (2009). Connected Vehicle Diagnostics and Prognostics, Concept, and Initial Practice. *IEEE Transactions on Reliability, 58*(2), 286-294.

Yu-Jen, Liu, Gary, W., Chang, E., et al., (2010). Mayr's equation-based model for pantograph arc of high-speed Railway traction system. *IEEE transactions on power delivery, 25*(3), 2025-2027.

Zhang, L. F., Xie, M. & Tang, L. C. (2007). A study of two estimation approaches for parameters of Weibull distribution based on WPP. *Reliability Engineering and System Safety, 92*, 360-368.

In: Case-Based Reasoning
Editor: Tao Lin
ISBN: 978-1-63482-952-6

Chapter 9

A STUDY OF APPLYING CASE-BASED REASONING ON DIAGNOSING AND MAINTENANCE

Yuo-Tern Tsai*

Department of Mechanical Engineering, De-Lin Institute of Technology, Taipei, Taiwan, ROC

ABSTRACT

Fault-diagnosis is a narrowing down procedure to identify the fault sources according to the symptoms of system broken down. Case based reasoning (CBR) is an efficient method in fault-diagnosis system (FDS) because it can cumulate and reuse the occurred maintenance experience. This chapter reported the methods of diagnosing and maintenance for mechanical systems based on CBR approaches. The methods of identifying fault resources and symptoms of a system are proposed for systematically clarifying the cause-result relationship between the faults and the symptoms. A logical process based on information flow analysis is proposed to quantify the cause-result relationship. The similarities of fault events are calculated by a quantified index for performing case matching in CBR diagnosing. The diagnostic system based on CBR diagnosing is constructed and subjected to an injection molding machine in cooperation with access database. The developed FDS is helpful for maintenance personnel in diagnosing and maintenance of injection molding machines.

Keywords: fault diagnosis, similarity, Case-Based Reasoning (CBR)

1. INTRODUCTION

Maintenance is more and more emphasized for modern engineering system to reduce production cost as well as promote business benefit. The shut-down time of an equipment should be kept as short as possible to meet the needs of effectiveness, especially for an

* Corresponding author: Department of Mechanical Engineering, De-Lin Institute of Technology, 1, Lane 380, Ching Yun Road, Tucheng, Taipei, Taiwan 236, ROC, Email: yttsai@dlit.edu.tw.

automated production system. The actions of a machine shut-down for maintenance usually involve checking, testing, repairing and/or replacing, etc. The possible activities of maintenance while a system is at "down" are shown in Figure 1. These activities each have a time interval associated with it that contributes to the length of the total down time. Among the time intervals, the diagnostic time may be the most significant one in the mean down-time (MDT) because it needs many professional techniques for identifying the problem points. To shorten the MDT, It is useful to develop a systematic method for diagnosing exactly what the problem is and how to fix it based on the collection and analysis of relevant data.

There are many diagnostic methods having been proposed for engineering systems based on various approaches. For example, a systematic approach on fault diagnosis of flexible manufacturing systems was proposed by Hu et al. [1] for integrating condition monitoring, fault identification and maintenance planning. The condition monitoring diagnostic methods have also been successfully applied in rotary machinery fault diagnosing [2]. To treat the uncertainty of the fault information, Tsai et al. [3] reported an evaluated method of system states and reliability based on a combination of Bayesian method and fuzzy signal information. Nguyen and Lee [4] employed a simple genetic algorithm (GA) to evaluate and select the optimized features for induction motor fault classification. The features were extracted from time vibration signals and used for the purpose of diagnosing motor faults. For complex systems, making use of modular, hierarchical architecture to explore the fault causes as well as fault diagnosis tools is a practical method.

Fault diagnosis is a narrowing down process from the system unit step by step into the component or element. The functional block diagram (FBD) of a system top to down can be traced as shown in Figure 2.

The overall plant (subsystem combination) forms the top level, the unit grade the second level and the parts/components the lowest level. Each block denotes a clearly identifiable function, such as a pump, a motor, a shaft, and so on. The diagnostic strategy is to narrow down the hierarchical items step by step based on the symptoms of the problem. The activities of fault diagnosis usually involve symptom definition, fault patterning, knowledge representation, faults inferring, and so on. The best strategy of developing fault-diagnosis system (FDS) is that the information of the fault data as well as available diagnostic knowledge can been completely integrated into the system. Ref. [5] addressed several aspects of CBM approach including definition, related international standards, procedure, and techniques, etc.

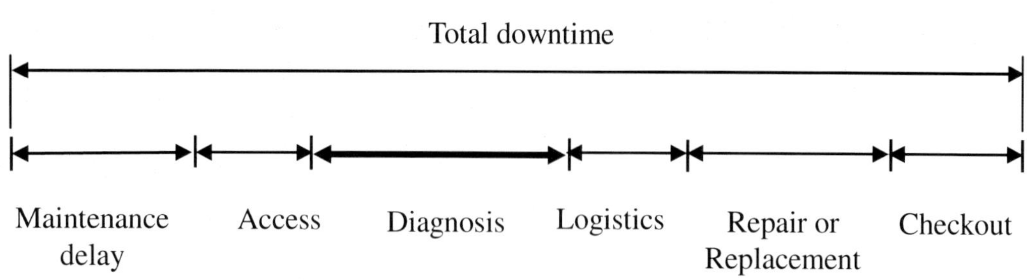

Figure 1. Downtime of Maintenance.

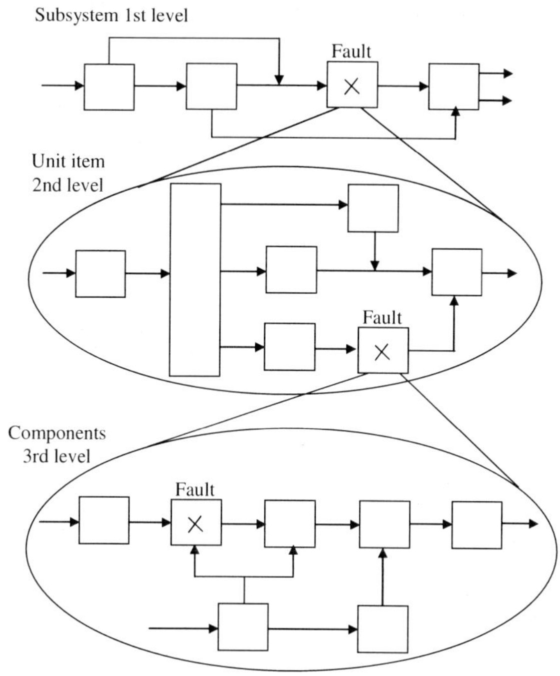

Figure 2. Hierarchical diagram of functional blocks.

Fault diagnostic system is usually developed depending on the types of knowledge representation and function requirements. Several commonly used tools in fault diagnosis are artificial neural networks (ANN), expert systems (ES) and case-based reasoning (CBR). For example, Su et al. [6] proposed a fault diagnosis assistance system where ES was used to construct a diagnostic knowledge base for motorcycle maintenance. Nahvil et al. [7] designed an ANN system for fault prediction in rotating machinery systems. Data from a vibration identification chart consisting of vibration signals for common rotating machinery faults were used to train the network. Wang and Hsu [8] proposed a Web-based CBR knowledge management system for PC troubleshooting based upon integrated cognitive task analysis and hierarchical clustering and ontology techniques. The properties of the ANN, ES and CBR are compared and listed in Table 1.

Focusing on CBR, the basic premise is that we can modify a strategy employed in a previously solved problem to solve a new problem. The CBR system supports the reuse and sharing of expert knowledge and problem solving. It can be used as an alternative to the traditional rule-based and model-based reasoning techniques. The problem-solving experience can be reused to solve the new problem. In the process of analogical problem-solving, the user first needs to receive and recognize the fault attributes. The knowledge system will feed similar past cases that match the terms back to the user based on these attributes. CBR has been successfully employed in a number of different types of problems such as in catering, recipe marking, dispute mediation and criminal sentencing [9]. The advantages of CBR retrieval compared with traditional key-term types of knowledge retrieval for fault diagnosis tasks mean that CBR retrieval could help to improve the performance of traditional fault-diagnosis methods [10]. An applications of CBR for image-based retrieval on polyurethane manufacturing was reported by Segura et al. [11].

Table 1. Comparisons of different diagnosis assistance systems

	Expert System	Case-Based Reasoning	Neural network
Data constructional method	If/then rules	Trouble-attributes retrieval	Recognition and fast pattern matching
Data request method	Expert thought process	Old cases	Learning algorithm and weights
How is it to be solved?	Logical step-by-step fashion	Find a previous similar solution	Recognizing the big picture
The degree of hardness-easiness to build	Difficult to develop in terms of knowledge acquisition	Easy to build and maintain but slower to diagnosis	Black box and lack of training data
Data renewal	By hand	By hand	Automatic learning
Is historical data enough?	No	Yes	Yes

CBR diagnosing can effectively reduce the training time of maintenance personnel, especially for a novice. It is carried out based on four steps, retrieving, reusing revising and retaining for reusing the previous maintenance experience. Here, CBR method is used to construct a fault diagnosis system (FDS) for injection molding machine (IMM). The similar studies has been presented in Ref.[12]. The possible faults and symptoms of the IMM are clarified by fault tree analysis and information flow analysis. A quantified method for calculating case similarity is proposed for systematizing the diagnostic algorithms of the CBR. The related pages of CBR diagnosing are designed in cooperation with an Access database. The results show that the FDS can provides a good support for maintenance of IMMs.

2. Fault Resources

Fault information includes fault causes and symptoms of a system at the various levels must first be established at first. An often used method to establish fault causes is by fault tree analysis (FTA). It is extensively applied to construct the hierarchical diagnostic model for maintenance in manufacturing systems [13].

2.1. Fault Tree analysis

Fault tree is a hierarchical structure of faults representing fault occurrence at different levels. It can be determined by Fault Tree Analysis (FTA). FTA is a logical deductive method for hierarchically establishing fault-causes in a system, from the overall system level down to the part/component level. The method is implemented based on the assumption that a fault can be located by analyzing the logical relationship between the system faults and its causes along a fault tree. It has the following characteristics:

(1) FTA can be used to deeply analyze a system fault level by level. It uses clear graphics to describe the internal valid logical relationships between the part/component faults and the system fault
(2) The fault tree clearly indicates which system fault is related to which part/component, what the relationship is, and how strong the relationship is. It also shows whether a part/component fault will cause a system fault, and if so, what and how great the effect will be.
(3) A fault tree provides a clear illustration for management and maintenance personnel, who may not have participated in the system design, to understand the system properties. This would help to shorten the training time of the maintenance personnel, and therefore cuts down the maintenance expense.
(4) Qualitative analysis using fault trees allows the designer to understand the fault modes and sources of the system, so as to find out weak links in the design scheme, and take corrective measures to remedy them.

Fault trees can be created following the hierarchical structure of the system functions from higher to lower. The hierarchical structure is based upon the principle that a system is comprised of subsystems, the subsystems are made up of modules, which in turn each have their own parts and/or components. Figure 3 shows the structure of a hierarchical system. The fault resources manifesting on the first level are related to the subsystems. The functional and sub-functional modules are related to the second and third levels of the faults, respectively. The parts/components are found at the lowest fault level. Each of the items in the hierarchy is not only the cause of an upper layer event but also the result of a lower layer event. Each corresponds to a real physical structure (based on field knowledge) or the embodiment of a principle.

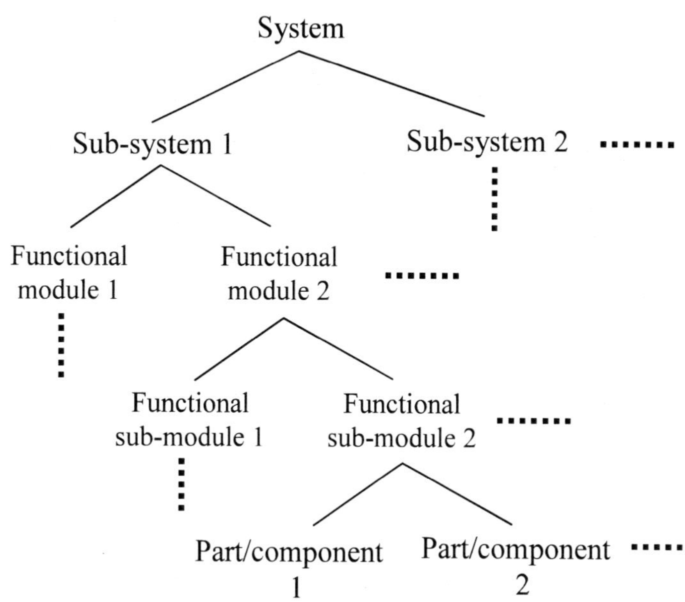

Figure 3. Hierarchical structure of a system.

The faults of a Machine can be classified into two categories sudden faults and gradually induced faults. The gradually induced faults are possibly occurred while machinery is running for a long time. This kind of faults possibly cause to a serious damage if the potential faults didn't be remedied in time. To avoid serious breakdown occurred, prognostic maintenance is commonly adopted to avoid unexpected failures. It is continuously motoring the performance states of systems so that the faults can be early detected and be solved in time [14].

2.2. Fault Identification

Fault symptoms are the basis of diagnosing since it is direct related to the fault points. Fault symptoms are usually unstructured and are not easily enumerated systematically so as to they are expressed in the form of linguistic descriptions for the state abnormality. A feasible approach to determine the symptoms of a system is to analyze the system's information flows. System information flows denote the information transformation among functional blocks and can be used to indicate the states of the corresponding function elements. System information flows include three main types, energy flow, material flow and signal flow [15]. The information flows deal with the following:

- Energy: mechanical, thermal, electrical, chemical, optical, nuclear, as well as force, current, heat, and so on;
- Material: gas, liquid, solid, dust, as well as also raw material, test samples, work pieces, end products, components, and so on;
- Signal: magnitude, display, control impulse, data, frequency, stability, and so on.

From the system information flows, the items for all function elements can be easily observed and measured, and fault symptoms determined accordingly. The FBDs of a system (from top to bottom) can be constructed according to the system information flows along with the comprehensive structure of function elements. A recent application of FBDs in fault diagnosing are proposed by the author for providing a helpful methodology for fault identification [16]. The diagnosis trees of the systems are constructed based on a functional block diagram consisting of the function elements and the test points.

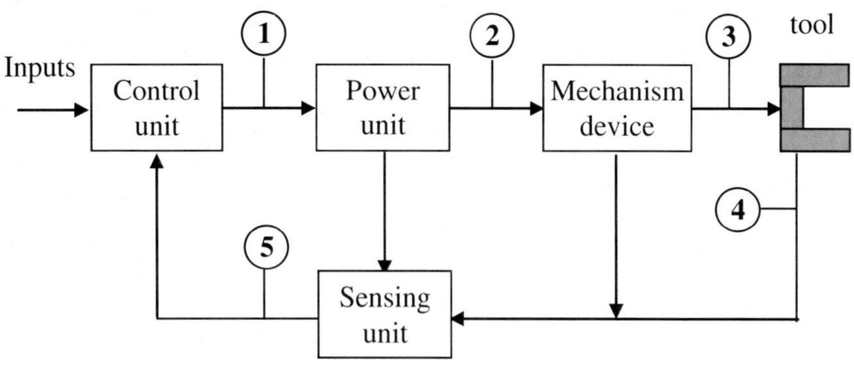

Figure 4. Functional block diagram of a mechatronic system.

Table 2. Fault patterns in a mechatronic system

Patterns	1 Control signals	2 Output energy	3 Moving velocity	4 Tool position	5 Signal display	Faults
1	N	N	N	N	N	Controller
2	Y	N	N	N	N	Power
3	Y	Y	N	N	N	Mechanism
4	Y	Y	Y	N	N	Tool
5	Y	Y	Y	Y	N	Sensor

The function elements of a mechatronic system, for example, may include the control unit, power device, mechanism, tools and sensors (F1-F5). The FBD for this system can be expressed as shown in Figure 4. There are five test points (T1-T5) in this FBD. The symptom attributes for the system are control signals, output energy, moving velocity, tool position, signal display. The deductive diagnosis can be carried out by inspecting the status of these test points. Considering the states of these test points, the fault patterns would be able to be set up as listed in Table 2. The diagnostic activities are to check the status of these test points. For example, a symptom set {Y, N, N, N, N}, where all test points flows are abnormal, except test point 1, indicates power unit failure. Possible fault sources can be identified by comparing the observed symptoms with the fault patterns. Further decomposition of the fault patterns at lower levels is necessary to finally identify the source of the fault.

This narrowing down procedure can be carried out systematically because the information flow of the function elements has a definite sequential relationship. This is because a definite correlation can be set up between the fault patterns and the symptoms. However, in some complex systems (such as mechanical systems), it is not always easy to clarify the sequential relationship of the information flow, due to the parallel or coupled properties of information flow. Ambiguity in the information relationship makes it difficult to narrow down the tasks.

3. CASE BASED REASONING (CBR)

The CBR system retrieves similar cases, and if necessary, adapts them to provide a desired solution when a new problem is given. The solution strategies of CBR are generated by modifying the solutions of the similar cases that are stored in the database according to the described problem structure. This method is easily accepted by users because the reasoning method is similar to the problem-solving strategy frequently used in human thought processes.

3.1. Diagnostic Problems

The fault-diagnosis domain is frequently characterized by ill-defined problems. In other words, clues relevant to solving diagnostic problems may not necessarily be described or understood simply by assessing their apparent characteristics. Cognitive biases, lack of knowledge and time pressure can all affect fault-diagnosis, and performance is often poor. Maintenance personnel often have a hard time in narrowing down the procedures during diagnosis due to the shortage of domain knowledge, especially for novices.

There is much evidence that novices have difficulty in troubleshooting and diagnosing. The reasons for maintenance novices being hard in diagnosing are listed in following (Schaafstal and Schraagen [17]).

(1) Lack of systematic thinking: Novices are not able to adopt functional thinking for problem solving and fault diagnosis. Broad searches of technical documents are therefore necessary to make up for this deficiency.
(2) Failure to apply analogical reasoning: Novices do not apply knowledge gained from previous experience to the problems they are currently facing, perhaps because they are not able to, or perhaps because they lack previous experience or knowledge to apply.
(3) Lack of systematic fault-diagnosis methods: Though details of relevant systems are available, novices cannot conduct fault diagnoses systematically.
(4) Failure to deduce similarities among problems: When a new problem is encountered, the similarity between the current and previous problems can only be deduced through systematic specifications and suggestions.
(5) Failure to identify fault symptoms correctly: Novices are often fuzzy about fault symptoms. Their memory loads may increase because of the symptom-masking phenomenon,, which prevents the application of fault-diagnosis strategies and weakens their performance.

A feasible approach to overcome some of the above cognitive problems is to use CBR for knowledge representation. It focuses on exploring the experiential knowledge from the past cases which can then be used to carry out fault diagnosis.

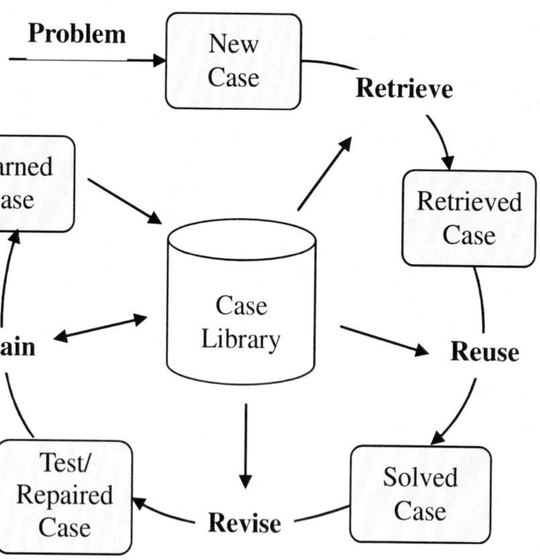

Figure 5. CBR cycle.

3.2. CBR Steps

CBR solves new problems by adapting successful solutions for similar problems. The appeal of CBR as a problem solving approach lies in its familiarity, where a solution will be obtained based on the solution of similar cases in the past. For example, doctors do not usually start each diagnostic procedure from the beginning. In most cases, they simply recall similar cases, where patients have had similar symptoms, and what treatments those patients received. The present treatment may be modified to meet the specific characteristics of the current patient, such as age, sex, weight and medical history.

The CBR process is illustrated in Figure 5 (Ref.[18]). The CBR is a cyclical process comprising the following actions:

(1) Retrieve the case(s) with the greatest similarity;
(2) Reuse the case solution to try to solve the problem;
(3) Revise the proposed solution if necessary;
(4) Retain the new solution.

A new problem (the target case) is matched against the cases in the case-base. The importance attached by the user to various features (indexes) of the case may be used to guide the matching process. One or more similar cases are retrieved from the case-base. A solution suggested for these cases is reused and tested for success. If necessary, the retrieved case(s) are revised to produce a new case, which can then be retained in the case-base.

A CBR system is generally composed of problem data, an action (or adjustment) and result data. When a maintenance engineer is confronted with a new problem he can acquire data for similar cases from the database. The cycle normally requires some user intervention. Most case adaptation in CBR systems is largely performed by the user. In other words, the CBR system acts primarily as an intelligent associative retrieval system. The following key issues need to be resolved for the construction of a feasible CBR system (Watson [19]):

(1) Case representation: how to structure cases and what case features should be stored;
(2) Indexing: how to assign indices to assist case retrieval;
(3) Retrieval: what retrieval algorithms should be used and what is meant by similarity? Techniques for dealing with similarity include nearest neighbor, induction and template retrieval;
(4) Adaptation: how can the cases be adapted to solve the current problem?

A fuzzy-CBR approach has been proposed to make the CBR system more user-friendly (Chan [20]). This approach advocates using linguistics variables in fuzzy theory to evaluate the case attributes. A linguistic variable is represented in a natural language form as well as by a fuzzy number. The textual description is intended to help users resolve uncertainty in their valuation of the case attributes. The fuzzy number representation and the associated fuzzy operators are used to calculate the similarity metrics for implementing the case-retrieving mechanism.

3.3. Case Similarities

The first steps of CBR diagnosis are to receive and recognize the fault attributes with the CBR knowledge expression for the analogical problem symptoms. The knowledge system of CBR will feed similar past cases that match the terms back to the user via the similarities of the these attributes.

A case similarity measure is employed in the retrieval process. Query cases are compared with cases stored in the case-base to find the most useful ones. The case-indexing procedure provides an efficient way to search for candidates. The search has a retrieval time which increases linearly with the number of cases and is most effective when the number of cases is fairly small. Users do not need to understand the relationship between the description and the solution since an automatic reasoning algorithm will feed back the proposed solutions.

The case-retrieval algorithm described in the study is mainly derived from the algorithm proposed by Kolodner [9]. The algorithm determines similarities between cases and identifies those with the greatest similarity values

$$\frac{\sum_{i=1}^{n} W_i \times sim(f_i^I, f_i^R)}{\sum_{i=1}^{n} W_i}, \tag{1}$$

where

n: number of attribute indices;
W_i: weighting value of the attribute index;
f_i^I: newly entered case;
f_i^R: cases in the case library;
sim (f_i^I, f_i^R): similarity between the newly entered case and the case in the case library;
i: attribute index variable;
I: identification mark for a newly entered case;
R: identification mark for an old case.

Determining a suitable set of weights is crucial to success. In the simplest form, users are simply presented with a list of key factors and asked to assign a numerical weight which reflects the significance they place on this attribute relative to the other attributes. These weights can then be used directly in the formula above to retrieve the nearest neighbors.

In this chapter, the weights are determined based on a correlation matrix describing the correlation between the causes and the symptoms of the faults. Assuming that a system has m causes and n symptoms, we construct a correlation matrix:

$$[a_{ij}] = \begin{bmatrix} a_{11} & a_{12} & \cdots & a_{1n} \\ a_{21} & a_{22} & \cdots & a_{2n} \\ \vdots & \vdots & \ddots & \vdots \\ a_{m1} & a_{m2} & \cdots & a_{mn} \end{bmatrix}, \tag{2}$$

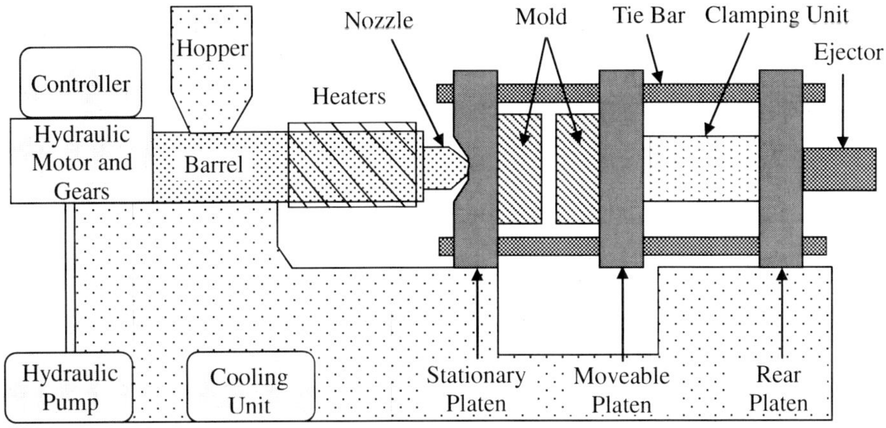

Figure 6. Injection Molding Machine.

where a_{ij} stands for the possibility of fault i occurring for symptom j. The possibility is quantified with 1, 3, 5, indicating Low, Medium, High, respectively. To obtain reasonable correlation values, we refer the FBD of system created by analyzing system information flows. However, the correlation assessment always necessitates more or less subjective judgments, due to cognitive differences. A feasible approach to obtain these values is by combining expert opinions using fuzzy evaluation. For general methods for pooling expert opinions using fuzzy operations.

Based on the correlation matrix, the weightings for various fault types can be defined by

$$W_{ij} = \frac{a_{ij}}{\sum_{j=1}^{n} a_{ij}} \quad . \tag{3}$$

Here, the weighting values change depending on the fault type. They characterize the symptom attributes for each fault, in accordance with the real fault symptoms.

4. Fault Analysis of IMMs

The CBR method is used to develop FDS for Injection Modeling Machines (IMMs) for helping maintenance personnel making a rapid diagnosis for the fault sources based on past maintenance records.

IMMs are used as a case to depicting CBR diagnosing. Normally, fault identification can be a time-consuming process, especially if this type of failure has not happened before or if the maintenance personnel are unfamiliar with the machinery. The IMM transfers the polymeric material (in powdered or granule form) from a feed hopper to a heated barrel. The resin is melted in the barrel and then injected into a mold. In this case, the resin is injected into the mold by a reciprocating screw or a ram injector. The reciprocating screw apparatus is placed inside the barrel. The reciprocating screw is able to inject a small amount of the total shot (amount of melted resin in the barrel). The mold for shaping the plastic is clamped shut under pressure with a platen arrangement and is held at a temperature well below the resin's melting point. The mold is cooled to a temperature that allows the resin to solidify and be cool to the touch. The mold plates are held together by hydraulic or mechanical force. The basic physical structure of the machine is shown in Figure 6 [12].

The possible faults of the machine are determined by FTA method. The decomposition is done according to the tree structure of the system. To simplify the diagnostic process, the fault level during CBR diagnosis is set to function module. This is to consider diagnostic complexity assessed by reviewing the historical maintenance records and face-to-face discussion with maintenance personnel. The possible fault types according to the problem classification for the IMM are listed as follows:

F1: Control unit failure;
F2: Power device failure;
F3: Injection mechanism failure;
F4: Clamping device failure;
F5: Ejection device failure;
F6: Cooling system failure;
F7: Mold unit failure.

The fault symptoms need to be given next. However, the symptoms are described using linguistic expressions which may be irregular and unstructured. To accurately recognize symptoms, the working procedures of the machine are reviewed. They are: (1)Mold closing and clamping; (2) Injection; (3) Holding under pressure and cooling; (4) Material dosing or metering; (5) Mold opening and part ejection. Based on the working procedures and the units for fault classification, the FBD of the machine is laid out, as shown in Figure 7.

In this example, there are eight test points set. The measurements include information on energy, material and signal flows. Reviewing the characteristics of the information flows, the symptom attributes for the machine are defined as follows:

S1: Controller status;
S2: Energy output;
S3: Injection activity;
S4: Mold opening/closing;
S5: Part ejection;
S6: Cooling system function;
S7: Mold temperature;
S8: Injection device temperature.

Discussing an example, provided a problem with the controller, such as being unable to start, signal output abnormalities, overload warnings, malfunctions of the safety door or safety rod, and so on. Abnormal conditions associated with these symptoms each are divided into three levels (0, 1, 2) representing (Normal, Slight and Critical), respectively. The levels are decided according to the complexity of maintenance needed to remedy the fault. For example, the abnormality of a fault which can be repaired by a field operator using only simple tools may be regarded as slight, but if the repair necessitates complicated tools, instruments and professional knowledge, it could be regarded as critical.

(1) Case knowledge

Each case in the case-library describes one particular situation, and all cases are independent of each other. The case structure includes four main parts: problem description, symptom states, fault sources (including causes), and solution strategy. An example of case for IMM maintenance is shown in Figure 8.

Now, a new problem must be solved. Several observations of the current situation and abnormalities caused by the new problem are observed. Note, not all feature values must be known. The observed symptoms of the new problem are shown in Figure 9.

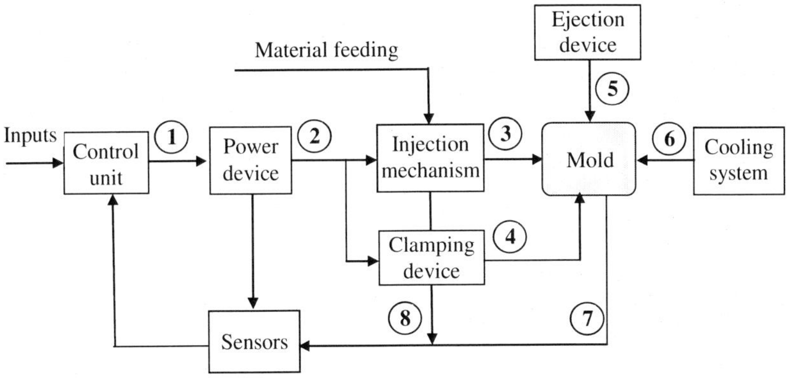

Figure 7. Functional block diagram of an IMM.

Case 1

Problem description:
The toggle mechanism for clamping does not work normally.
Symptoms:

- Controller state: Normal
- Energy output: Critical abnormality
- Injection activity: Slight abnormality
- Mold opening/closing: Slight abnormality
- Part ejection: Slight abnormality
- Cooling system function: Normal
- Mold temperature: Normal
- Injection device temperature: Normal

Fault sources: Power device

- Causes: The toggle filler oil is inadequate

Solution: Add hydraulic oil

Figure 8. Case example.

Problem description:
Material feeding screw cannot be driven.
Symptoms:

- Energy output: Critical abnormality
- Injection activity: Slight abnormality

Figure 9. The symptoms of a new problem.

Here, the other unlisted values of the symptoms are at the defeated value (Normal).

(2) Case Similarities

The correlation matrix for this example is

$$[a_{ij}] = \begin{bmatrix} 5 & 3 & 3 & 1 & 1 & 1 & 5 & 5 \\ 3 & 5 & 3 & 3 & 1 & 3 & 1 & 1 \\ 3 & 3 & 5 & 1 & 1 & 1 & 1 & 3 \\ 1 & 3 & 1 & 5 & 3 & 1 & 1 & 1 \\ 1 & 3 & 1 & 1 & 5 & 1 & 3 & 1 \\ 3 & 3 & 1 & 1 & 1 & 5 & 3 & 1 \\ 1 & 1 & 3 & 1 & 3 & 3 & 5 & 1 \end{bmatrix} \quad (4)$$

The weights for the various faults can be obtained with Eq.(2). The case similarity can be calculated by revising Eq.(1) as follows:

$$\sum_{j=1}^{8} W_{ij} \times \left(1 - \frac{s_j^I - s_j^R}{2}\right), \quad (5)$$

where so is the assessed level of the symptoms. For example, the symptoms for the new problem for case 1 are {0, 2, 1, 0, 0, 0, 0, 0} and {0, 2, 1, 1, 1, 0, 0, 0}. The resultant similarity between the two cases would be 0.9. Here, the weight set W2j{0.15, 0.25, 0.15, 0.15, 0.05, 0.15, 0.05, 0.05} is chosen since the fault type of case 1 is F2.

The most similar case is chosen in the solution reuse process and this problem solution adapted. The new fault problem is then reviewed by discussing it with the equipment operator. If the cause of the problem is judged to be a burnt-out drive motor the solution is to replace the motor. If this diagnosis proves to be correct, the solution and the new problem form a new case which is stored in the case library.

5. Fault-Diagnosis System (FDS)

The FDS was constructed including two functions for diagnosing and maintenance of an IMMs. One is case management of the case library which includes four operations: adding, deleting, modifying and inquiring. The other one is case-reasoning, where the most similar cases are found, and the case solution reused for the new problem. The related pages of diagnosing were constructed using the ASP programming language. The original IMM maintenance data, including the problems and the solution strategies, were provided by a factory. The data format was revised to conform to the case structure so that case similarity could be computed for progressive case retrieval. The cases were stored in a Microsoft Access database. A page including four functions adding new, deleting, modifying and querying is designed for case management

Several diagnostic steps and displays are illustrated in the figures.

(1) Users input the problem descriptions and the abnormal levels associated with the new problem. The interface of diagnosing is shown in Figure 10. In this example, the new problems are "The material feeding screw cannot be driven". The appeared symptoms are s2 (critical) and s3 (slight).
(2) The CBR diagnostic results are shown in Figure 11. The system automatically computes the case similarity and sorts them in descending order of similarity. In this example, the most similar case would be case 1 which forms the basis of the new problem solution.
(3) The problem symptoms and the solution strategies are shown in Figure 12. The page offers the function of modification for the symptoms and solutions according to the retrieval case. It is automatically generated by pressing the adaptation button.

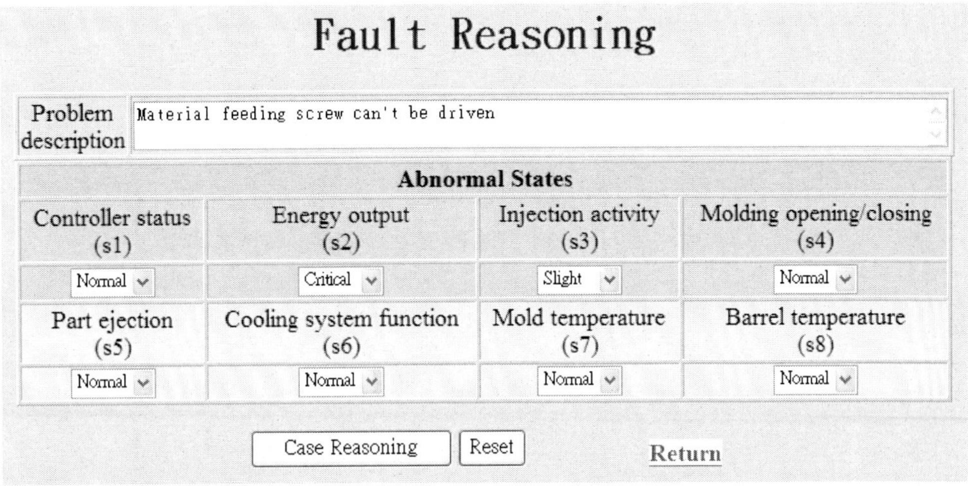

Figure 10. Case retrieval page.

Diagnostic results

The Problem is: Material feeding screw can't be driven (s1~s8) { 0 2 1 0 0 0 0 0 } **Adaption** **Return**

ID	dat	line	machineID	s1	s2	s3	s4	s5	s6	s7	s8	Fault	Problems	Fault sources	Solution	Time (hr)	similarity
1	2007/1/12	PL405	DB1251-05 01 XVM4316	0	2	1	1	1	0	0	0	2	The toggle mechanism of clamping doesn' t work normally.	Power device， The toggle filler oil inadequate.	Supplement hydraulic oil	0.5	0.9
3	2007/1/13	PL420	DB0895-01 01XVM2916	1	1	2	0	0	0	0	0	3	Material can't be dosed into mold normally	Injection mechanism, Injection nozzle jamming	Remove materials attached on the leadwire and clean the nozzle	1.5	0.72
6	2007/1/30	PL412	DB1251-01 01VM3916	1	0	1	0	0	0	0	2	3	Injection machine can't be heated normally	The fuse of barrel heater burned out	Replace the heater fuse	0.5	0.605
2	2007/1/12	PL438	VM123	1	0	1	0	0	0	0	2	3	The working temperature of injection is too low.	Injection mechanism, The heater burned out	Replacement the heater	2	0.605

Figure 11. CBR diagnostic results.

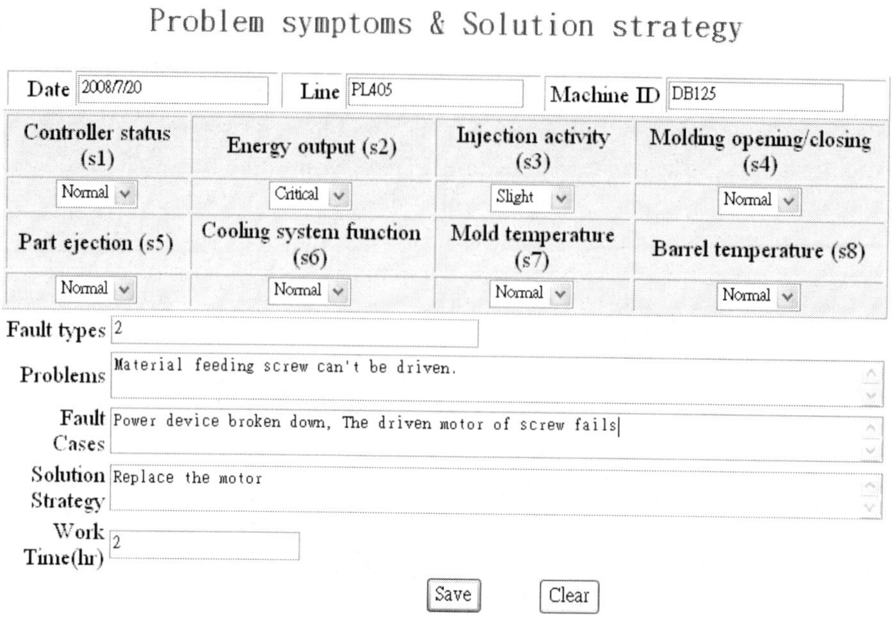

Figure 12. Case revision.

ID	dat	line	machineID	s1	s2	s3	s4	s5	s6	s7	s8	Fault	Problems	Fault sources	Solution	Time (hr)
1	2007/1/12	PL405	DB1251-05 01 XVM4316	0	2	1	1	1	0	0	0	2	The toggle mechanism of clamping doesn' t work normally.	Power device， The toggle filler oil inadequate.	Supplement hydraulic oil	0.5
2	2007/1/12	PL438	VM123	1	0	1	0	0	0	0	2	3	The working temperature of injection is too low.	Injection mechanism, The heater burned out	Replacement the heater	2
3	2007/1/13	PL420	DB0895-01 01XVM2916	1	1	2	0	0	0	0	0	3	Material can't be dosed into mold normally	Injection mechanism, Injection nozzle jamming	Remove materials attached on the leadwire and clean the nozzle	1.5
4	2007/1/16	PL403	DB125	2	0	0	0	0	1	1	1	1	Controller unstable, Pannel functions occasionally are disabled	Control software abnormality, Unexpected shut-down occurring	Reinstall control software	2.5
5	2007/1/29	PL	DB125	1	0	0	0	0	2	2	0	6	Mold temperature variety irregularity	Cooling warter temperature too high, The heat sink of cooling system fails	Replace the heat sink	1.5
6	2007/1/30	PL412	DB1251-01 01VM3916	1	0	1	0	0	0	0	2	3	Injection machine can't be heated normally	The fuse of barrel heater burned out	Replace the heater fuse	0.5
7	2008/7/20	PL405	DB125	0	2	1	0	0	0	0	0	2	Material feeding screw can,t be deriven	Power device broken down. The driven motor of screw fails	Replace the motor	2

Figure 13. Case retained.

(4) The adapted solution, together with the diagnostic problem, form a new case, which is stored in the case library (the 7th case in Figure 13).

CONCLUSION

Fault diagnosing is largely dependent upon the practical knowledge and experience of the maintenance personnel. Making use of past maintenance experience is an efficient approach to shorten the maintenance time. The study reported a FDS based on CBR to perform fault diagnosis for maintenance of IMMs. The fault types and symptoms are derived by the FTA method, which are used to compile an index for retrieving similar cases. The relative importance of the symptom attributes is evaluated based on a symptom to cause correlation matrix. The advantages of the FDS are that similar cases can be retrieved quickly simply by inputting a set of abnormal levels (of symptom attributes) and the solution can be generated accordingly. The CBR diagnosis could be easily applied to other types of equipment just by modifying the attributes of faults and symptoms. The system will become a more and more powerful because the case library is continuously being enlarged depending on the used time increased.

The fault sources and symptoms are the basis of diagnosing. The symptom attributes and the symptom-cause correlations are easily set up by analyzing system information flows. CBR is a technique for solving problems based on past experience by four phases: retrieve, reuse, revise and retain. Fault diagnosis based on CBR can assist in treating ill-defined problems associated with the diagnostic activities. It offers a direct and rapid method for diagnosing faults and generating solutions, and can efficiently simplify the maintenance of a complex system. The weights of symptom attributes have a big effect on the similar cases retrieved and the diagnostic results as well as defining symptom attributes appropriately is crucial in retrieving similar cases and insuring that the proper solution is retrieved.

REFERENCES

[1] Hu, W; Starr, AG; Zhou, Z. Leung, AYT. A systematic approach to integrated fault diagnosis of flexible manufacturing systems, *International Journal of Machine Tools and Manufacture*, Vol. 40(11), 1587-1602, 2000.

[2] Zio, E; Baraldi, P; Gola, G. Ensemble feature selection for diagnosing multiple faults in rotating machinery, *Proc. IMechE, Part O: Journal of Risk and Reliability*, Vol. 221(1), 29-41, 2008.

[3] Tsai, YT; Wang, KS; Lin, HY. "Performance states estimation of system using Bayesian method and fuzzy signal information", *Journal of Chinese Society of Mechanical Engineers*, Vol.31, No.5, 403~411, 2010.

[4] Nguyen, NT; Lee, HH. Improvement of induction motor fault diagnosis performance by using genetic algorithm-based feature selection, *Proc. IMechE, Part C: J. Mechanical Engineering Science*, Vol.222(8), 1613-1619, 2008.

[5] Shina, JH; Junb, HB. "On condition based maintenance policy", *Journal of Computational Design and Engineering*, Vol.2, Iss 2, 119–127, 2015.

[6] Su, KW; Hwang, SL; Chou, YF. Applying knowledge structure to the usable fault diagnosis assistance system: A case study of motorcycle maintenance in Taiwan, *Expert Systems with Application*, Vol.31, 370-382, 2006.

[7] Nahvil, H; Esfahanianl, M. Fault identification in rotating machinery using artificial neural networks, *Proc. IMechE, Part C: J. Mechanical Engineering Science*, Vol.219(2), pp.141-158, 2005.

[8] Wang, SL; Hsu, SH; A web-based CBR knowledge management system for PC troubleshooting, *International Journal of Advanced Manufacturing Technology*, Vol.23, 532-540, 2004.

[9] Kolodner, JL. Cased-Based Reasoning, Morgan Kaufmann Publishers, San Mateo, 1993.

[10] Hsu, SH; Wang, SL; Evaluating the effectiveness of knowledge-retrieval aids for fault-diagnosis tasks, *Journal of The Chinese Institute of Industrial Engineers*, Vol.21(5), 465-474, 2004.

[11] Segura Velandia, DM; West, AA. Image-based retrieval in case-based reasoning systems for polyurethane manufacture, *Proc. IMechE, Part B: Journal of Engineering Manufacture*, Vol.223(1), 089-105, 2009.

[12] Tsai, YT. "Applying a case-based reasoning method for fault diagnosis during maintenance", *Proc. IMechE, Part C:* J. Mechanical Engineering Science, Vol.223, 2431-2441, 2009.

[13] Hu, W; Starr, AG; Leung, AYT. Operational fault diagnosis of manufacturing systems, *Journal of Materials Processing Technology*, Vol.133, 108-117, 2003.

[14] Tsai, YT; Chen, CC; Liao, ZX. "A study of performance degradation prediction and fault diagnosis for mechanical systems", *Journal of Technology*, Vol.27/3, 121-129, 2012.

[15] Tsai, YT; Wang, KS. The development of modular-based design in considering technology complexity, *European Journal of Operational Research*, Vol.119(3), 138-149, 1999.

[16] Tsai, YT; Hsu, YY. "A study of function-based diagnosis strategy and testability analysis for a system", *Proc. IMechE, Part C: J. Mechanical Engineering Science*, Vol.226, 273-282, 2012.

[17] Schaafstal, A; Schraagen, JM. Training of troubleshooting: A structured, task analytical approach, Cognitive Task Analysis, Lawrence Erlbaum Associates, Publishers, London: Mahwah, New Jersey, 57-70, 2000.

[18] Aamodt, A; Plaza, E. Case Based Reasoning: Foundational Issues, Methodological Variations and System Approaches, *AI Communications*, Vol.7(1), 39-39, 1994.

[19] Watson, I. Applying Case-Based Reasoning: Techniques for Enterprise systems, Morgan Kaufmann, 1997.

[20] Chan, FTS. Application of a hybrid case-base reasoning approach in electroplating industry, *Expert Systems with Application*, Vol.29, 121–130, 2005.

INDEX

H

I

K

L

M

O

P

Q

R

S

T

U

V

W